Practical Geolocation for Electronic Warfare Using MATLAB®

Practical Geolocation for Electronic Warfare Using MATLAB®

Nicholas A. O'Donoughue

ARTECH HOUSE

BOSTON | LONDON
artechhouse.com

Library of Congress Cataloging-in-Publication Data
A catalog record for this book is available from the U.S. Library of Congress.

British Library Cataloguing in Publication Data
A catalogue record for this book is available from the British Library.

Cover design by Robert Oakley

Ebook ISBN: 978-1-63081-889-0
Print ISBN: 978-1-63081-888-3

685 Canton Street
Norwood, MA 02062

Software:
https://github.com/nodonoughue/emitter-detection-book

Videos:
https://us.artechhouse.com/Assets/bkvdo/odo_888/1-1.html
https://us.artechhouse.com/Assets/bkvdo/odo_888/1-2.html
https://us.artechhouse.com/Assets/bkvdo/odo_888/2-1.html
https://us.artechhouse.com/Assets/bkvdo/odo_888/2-2.html
https://us.artechhouse.com/Assets/bkvdo/odo_888/2-3.html
https://us.artechhouse.com/Assets/bkvdo/odo_888/3-1.html
https://us.artechhouse.com/Assets/bkvdo/odo_888/3-2.html
https://us.artechhouse.com/Assets/bkvdo/odo_888/4-1.html
https://us.artechhouse.com/Assets/bkvdo/odo_888/4-2.html
https://us.artechhouse.com/Assets/bkvdo/odo_888/5-1.html
https://us.artechhouse.com/Assets/bkvdo/odo_888/5-2.html
https://us.artechhouse.com/Assets/bkvdo/odo_888/6-1.html
https://us.artechhouse.com/Assets/bkvdo/odo_888/6-2.html
https://us.artechhouse.com/Assets/bkvdo/odo_888/7-1.html
https://us.artechhouse.com/Assets/bkvdo/odo_888/7-2.html
https://us.artechhouse.com/Assets/bkvdo/odo_888/8-1.html
https://us.artechhouse.com/Assets/bkvdo/odo_888/8-2.html

10 9 8 7 6 5 4 3 2 1

To my eternally patient wife

Contents

Preface

This text explores many (but not all!) of the practical realities that arise from the employment of geolocation for electronic warfare in real-world systems. These realities include the impact of a priori knowledge about the position of the target, errors in sensor position, orientation, or velocity, and the impact of repeated measurements over time.

We build upon the derivations and source code from *Emitter Detection and Geolocation for Electronic Warfare*, published in 2019. Where it is necessary, material from that text is included or summarized here, but it is recommended that readers first familiarize themselves with that text.

The companion software for both books can be downloaded from the publicly available Git repository.

The intended audience for this book includes engineers and electronic warfare practitioners who are interested in the development or employment of geolocation algorithms. As with my last text, the MATLAB® code is intended to provide a starting point for the development of novel algorithms in a way that allows for comparison to classical approaches.

Chapter 1

Introduction

Control of the electromagnetic spectrum (EMS) is critical for success on the modern battlefield. In 2020, then-Secretary of Defense, Hon. Mark Esper, wrote "These challenges [from peer and near-peer adversaries] have exposed the cross-cutting reliance of U.S. forces on the EMS" [1]. Later in that document, he introduced a significant organizational change in how the United States will approach EMS Operations (EMSO). These doctrinal and policy changes are not the focus of this text, but they do set the tone for the importance of EW, as a principal component of modern military operations.

In this text, we will focus on a subset of EW: the geolocation of sources of electromagnetic energy. What sets this apart from previous texts, such as *Emitter Detection and Geolocation for Electronic Warfare* [2], is our inclusion of realistic constraints and sources of uncertainty, which we will briefly describe in this chapter.

1.1 RECEIVER PROCESSING

We will not discuss in detail the reception of signals or their processing to detect transmissions of interest and to estimate key parameters. For clarity, this process is briefly reviewed here.

A sample receiver block diagram is shown in Figure 1.1. Note that there are many receiver architectures, with more or fewer components, than are shown here. This diagram is not meant to be definitive or exhaustive, but illustrative. It begins at the left edge with an aperture, followed by a series of filters and mixers that isolate the band of interest and convert it to a frequency that can be sampled. After digitization, the signal is fed through a detector to identify the presence of signals of interest, and a pulse parameter estimation block, to estimate critical pulse parameters, such as center frequency, chirp rate, and modulation type. At this point, most EW systems generate a *pulse descriptor word* (PDW) to describe the intercepted pulse.

The next step, typically, is to associate PDWs over time, in order to keep track of an emitter from pulse to pulse, and to estimate additional parameters, such as the

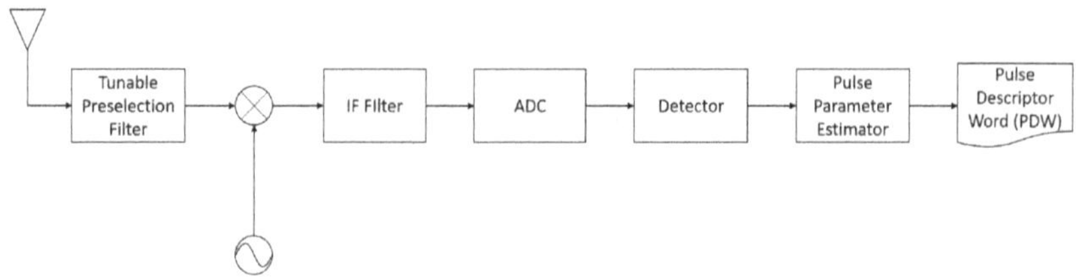

Figure 1.1 Block diagram of a notional receiver.

pulse repetition interval (PRI), duty cycle, or frequency hopping characteristics. At this point, a classifier may attempt to interpret the type of signal (communications, weather sensing, surveillance radar, etc.).

We will assume in this text that all of these steps have occurred already and that we are presented with some set of measurements (ζ) of underlying signal parameters ($\mathbf{z}$); our task is to perform geolocation of the source of those emissions to determine the coordinates ($\mathbf{x}$) from which they were transmitted.

For an in-depth discussion of receivers, including specific hardware choices and signal conditioning, interested readers are referred to Poisel's *Electronic Warfare Receivers and Receiving Systems* [3]. For a discussion of the detection and parameter estimation steps, see Parts I and II of *Emitter Detection and Geolocation for Electronic Warfare* [2], as well as Chapter 6 of [4], and Chapters 2 (for detection of standard communications signals) and 4 (for signals with low probability of intercept features, such as frequency hopping) of [5]. For a more theoretical and general discussion of detection and estimation theory, see [6], [7].

1.2 GEOLOCATION

Geolocation can be simplified to the process of estimation of a source's position $\mathbf{x}$, given some set of observable parameters $\mathbf{z}$, and a equation $f()$ that relates them:[1.]

$$\mathbf{z} = f(\mathbf{x}) \tag{1.1}$$

This process is more difficult than a simple inverse problem because the parameters $\mathbf{z}$ are not sampled directly, rather we are presented with a stochastic variable ζ that varies with the parameters $\mathbf{z}$ as well as random fluctuations, such as those due to measurement noise. In many cases, we can approximate ζ with a Gaussian random variable distributed:

$$\zeta \sim \mathcal{N}(\mathbf{z}, \mathbf{C_z}) \tag{1.2}$$

1. This is a simplified form of the measurement model $f()$, which must by necessity also consider the position of the sensors ($\mathbf{x}_n, n = 0, \ldots, N-1$), and possibly additional parameters, such as their velocities (for FDOA).

where the expectation of ζ is $\mathbf{z}$, but the covariance matrix $\mathbf{C_z}$ represents the fluctuations.[2] These statistical underpinnings of target geolocation are reviewed in Chapter 2.

In Part III of *Emitter Detection and Geolocation for Electronic Warfare* [2], we reviewed three common phenomenologies for geolocation: angle of arrival (or triangulation), time difference of arrival (TDOA), and frequency difference of arrival (FDOA), illustrated in Figure 1.2.

Figure 1.2(a) illustrates AOA geolocation via the crossing of three lines of bearing that intersect near the source of a signal. In this case, the measurements are the direction of arrival of the signal at each sensor. Two sensors are required to form a solution in a 2-D space, and three sensors are required for a 3-D solution. Algorithms to solve for geolocation estimates via AOA can be found in the `+triang` folder of the provided MATLAB® source code.

Figure 1.2(b) illustrates TDOA geolocation via the use of time of arrival estimates for a signal at three stations, which are compared to generate two *isochrones*, that represent lines of constant time *difference* of arrival between two pairs of stations. These isochrones intersect at the estimated location of the signal's source. In this manner, three stations are required for a 2-D solution, but four are required for 3-D. Algorithms to solve for geolocation estimates via TDOA can be found in the `+tdoa` folder of the provided MATLAB® source code.

Figure 1.2(c) illustrates FDOA geolocation via the use of frequency estimates for a signal at three stations, which are compared to generate two *isodoppler contours* that represent lines of constant frequency *difference* of arrival between two pairs of stations. Since FDOA is reliant not just on where the source and sensors are, but also on their relative velocities, the contours are much more complex than in AOA or TDOA. Algorithms to solve for geolocation estimates via FDOA can be found in the `+fdoa` folder of the provided MATLAB® source code.

Any of these techniques can be combined for improved stability, or reliable 3-D geolocation with fewer sensors. Algorithms to solve for geolocation estimates via hybrid AOA/TDOA/FDOA can be found in the `+hybrid` folder of the provided MATLAB® source code.

1.2.1 Practical Considerations of Geolocation

While the algorithms presented in [2] provide working solutions for AOA, TDOA, and FDOA geolocation, they rely on a set of simplifying assumptions for clarity. These assumptions are: (1) that all target positions are equally likely, (2) that all sensors have perfect knowledge of their position, orientation, and velocity, and (3) that all measurements are simultaneous (or near enough to simultaneous that the target is

2. It may be there is also a *bias* term $\mathbf{b}$ in our measurements, in which case the expectation would be given $\mathbf{z} + \mathbf{b}$. It's also possible that the expectation could be a nonlinear function of the true parameter. We will assume in this text that the measurement vector ζ has been properly processed such that any bias or nonlinearities are removed and its expectation is $\mathbf{z}$.

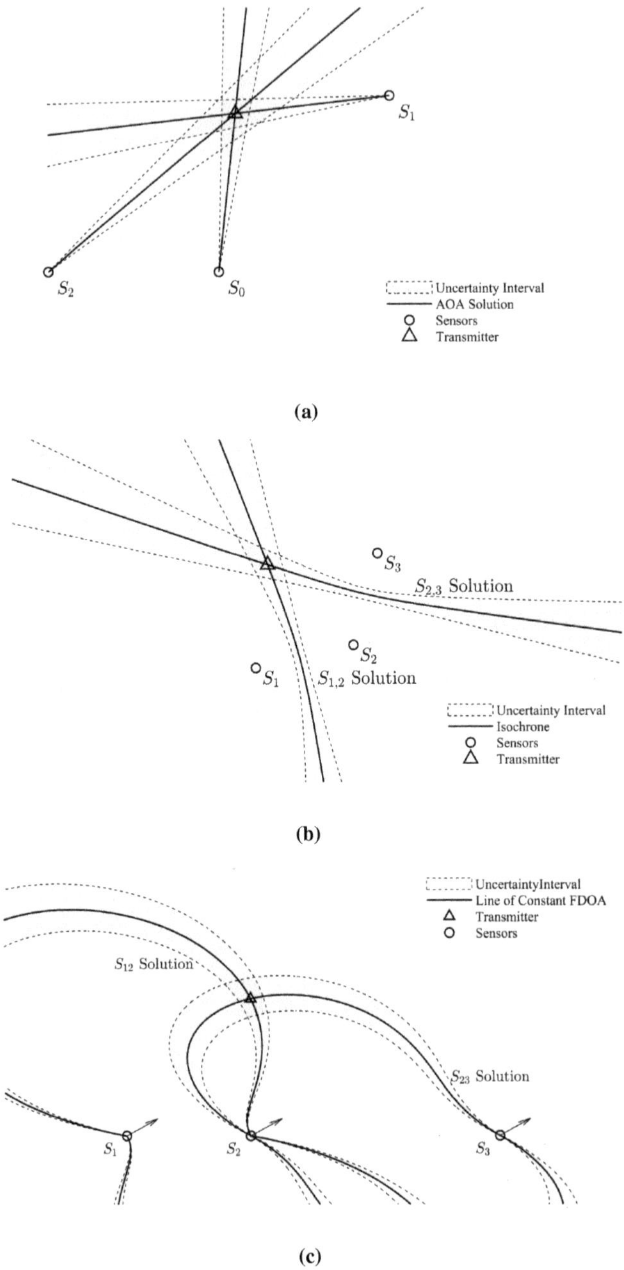

Figure 1.2 Common methods for solving geolocation, discussed in [2]. (a) AOA, (b) TDOA, and (c) FDOA.

assumed to be stationary). We now briefly discuss each, and how they will be relaxed in this text.

1.2.1.1 Target Constraints

In our previous text, we assumed that all possible geolocation solutions were equally likely. In other words, there was no a priori information on target location. While this assumption is reasonable algorithmically, it is often not operationally reasonable. This can be violated in two ways: hard constraints and soft constraints.

Hard constraints take the form of fixed equality or inequality conditions, such as knowledge that the target is a maritime vessel and must therefore be on the Earth's surface, or that an aircraft is traveling at a known altitude. Hard constraints are expressed in the form

$$\begin{array}{ll} \mathbf{a}^T\mathbf{x} = 0 & \text{Equality constraint} \\ \mathbf{b}^T\mathbf{x} \leq 0 & \text{Inequality constraint} \end{array} \tag{1.3}$$

If there are multiple constraints, we can express them concisely in matrix/vector notation:

$$\begin{array}{ll} \mathbf{A}\mathbf{x} = \mathbf{0} & \text{Equality constraints} \\ \mathbf{B}\mathbf{x} \leq \mathbf{0} & \text{Inequality constraints} \end{array} \tag{1.4}$$

If the constraints are nonlinear, then we write them:

$$\begin{array}{ll} \mathbf{a}(\mathbf{x}) = \mathbf{0} & \text{Nonlinear equality constraints} \\ \mathbf{b}(\mathbf{x}) \leq \mathbf{0} & \text{Nonlinear inequality constraints} \end{array} \tag{1.5}$$

It is straightforward to implement a constrained solver that considers this information, particularly with linear constraints, which are convex [8].[3] Predicting estimation performance is slightly more difficult; we will primarily use the *constrained* Cramér-Rao lower bound (CRLB) to predict performance [9]–[11].

Soft constraints present statistical information about where the target is or is not likely to be located. Soft constraints can come from external measurements, such as a separate system's estimated target position, or prior estimates of a target's position and velocity (e.g., from a tracker). Soft constraints are represented as a prior probability distribution function $f_{\mathbf{X}}(\mathbf{x})$. In this manner, the *posterior* likelihood function for any set of measurements ζ is given:

$$\ell(\zeta, \mathbf{x}) = \ell(\zeta|\mathbf{x}) f_{\mathbf{X}}(\mathbf{x}). \tag{1.6}$$

where $\ell(\zeta|\mathbf{x})$ is the *likelihood function*, which computes how likely it is that the measurement ζ would occur if the true target location were $\mathbf{x}$. The *posterior* likelihood

3. Note that for convex optimization methods to be used, both the constraints and the *cost function* (in this case, the likelihood function or a simplified form of it) must also be convex. This is generally not true with geolocation problems, but they can often be assumed to be locally convex in a region around the true location $\mathbf{x}$.

function $\ell(\zeta, \mathbf{x})$ reflects the combination of the measurement likelihood function and the *a priori* probability distribution function $f_{\mathbf{X}}(\mathbf{x})$.

In Chapter 5, we will discuss constraints, solving them, and their impact on performance.

1.2.1.2 Sensor Uncertainties

Traditionally, sensors that engage in geolocation of adversary transmitters have been placed on fixed (permanent, or semimobile) towers, satellites with well-known trajectories, or even wide-bodied aircraft (such as the RC-135 Rivet Joint used by the U.S. Air Force, shown in Figure 1.3(a)). In these scenarios, it is often reasonable to assume that the sensor's position, orientation, and velocity are known accurately.

However, these sensors are increasingly miniaturized, and the modern battlefield may include their use on smaller platforms that operate in a heavily contested electromagnetic spectrum, such as the Multi-Function Electronic Warfare–Air Large (MFEW-AL) Army program, to be fielded on the MQ-1C Gray Eagle (an artist's rendition is shown in Figure 1.3(b)), or the backpack sensor being fielded by the Army for use by dismounted soldiers: the Versatile Radio Observation and Direction (VROD) system (shown in Figure 1.3(c)). The end result is that ESM sensors are (a) more susceptible to turbulent airflow (and, in the case of ground vehicles or dismounted soldiers, uneven terrain), (b) constrained in the quality of their navigational sensors (due to limited size, weight, and power, and lower cost constraints), and (c) potentially unable to utilize precision navigation and timing services, such as GPS (due to proximity to hostile forces), resulting in increased sensor position errors.

Thus, it is necessary to determine the impact of errors in not only a sensor's position, but also its orientation (for AOA), velocity (for FDOA), and drift in both the clock (for TDOA) and local oscillator (for FDOA). In Chapter 6, we will discuss these errors, and formulate their impact to the geolocation problem via their impact to the received signal vector ζ.

1.2.1.3 Repeated Measurements

For fixed or stationary targets, the use of repeated measurements over time results in high accuracy, even if the individual sensor updates have poor accuracy, as long as they are *unbiased*. This process is illustrated in the iterative estimation block diagram of Figure 1.4.

In Chapter 7, we will examine an optimal solution, in which all prior measurements are stored, and the estimate is recomputed at each update interval using all available information. This will then be compared with an iterative approach that matches Figure 1.4, wherein an estimate of the transmitter's position (along with expected error covariance for that estimate) is propagated forward and combined with the new sensor measurements. We will show the sufficiency of iteratively updating measurements.

(a) (b)

(c)

Figure 1.3 Example ESM sensors. (a) RC-135U Combat Sent and RC-135U/V Rivet Joint (the aircraft designation depends on mission configuration). © 2012 U.S. Air Force[12]. (b) Artist rendering of MQ-1C Gray Eagle UAV with a proposed MFEW-Air Large sensor pod. © 2019 Lockheed Martin[13]. (c) Soldier with VROD backpack sensor. © 2019 U.S. Army[14].

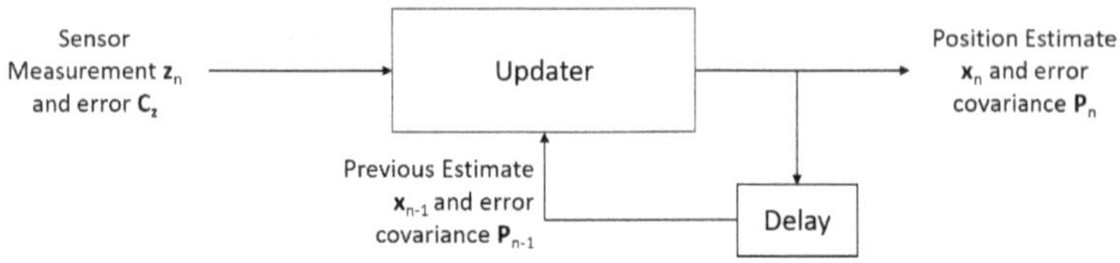

Figure 1.4 Block diagram of iterative estimation.

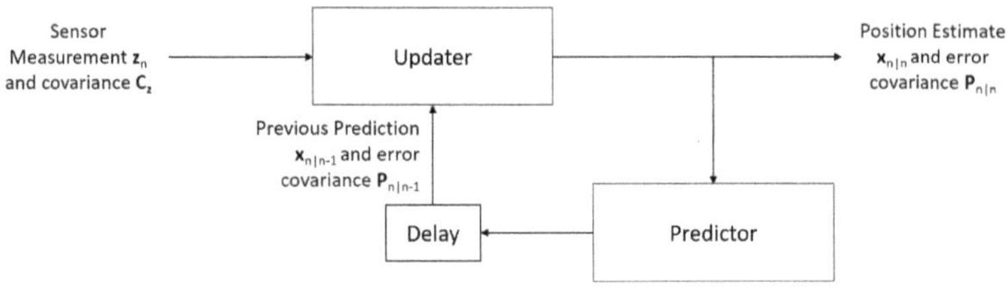

Figure 1.5 Block diagram of tracker predict-update cycle.

1.2.1.4 Tracking

For mobile targets, iterative estimation is replaced with tracking, since the underlying position $\mathbf{x}$ is no longer constant (replaced with $\mathbf{x}(t)$). A basic tracker loop consists of an iterative process of first updating the estimate of the underlying target state $\mathbf{x}_{n|n}$ (in this case, position and, potentially, velocity and acceleration terms) to consider the latest sensor measurement (ζ_n), and then prediction of what that state will be when the next sensor measurement is taken ($\mathbf{x}_{n+1|n}$).[4] When the next measurement is available, the process repeats, as shown in Figure 1.5.

Much can be written about tracking; indeed countless textbooks have been, and so we will only scratch the surface here.

In Chapter 8, we will introduce trackers with a discussion of the basic $\alpha-\beta$ and $\alpha-\beta-\gamma$ filters and the Kalman Filter, along with common target motion models, and relevant measurement models. We will briefly discuss issues around convergence of the tracker and prediction of steady-state performance. Deployed trackers must contend with a complex problem of data association (assigning measurements to new or existing tracks), track management (removing, or *pruning*, tracks that have gone stale), complex models (e.g., motion models that represent different target states, such as level flight vs. turning), bias reduction, accuracy prediction, and confidence scoring. Most of these topics are beyond the scope of this report, but interested readers are referred to [15] for a good detailed text on the formulation and execution of trackers, including detailed discussion of data association, as well as [16], which focuses on challenges associated

4. The subscript $\cdot_{n+1|n}$ indicates that the prediction is for time index $n+1$ but only considers information available at time index n.

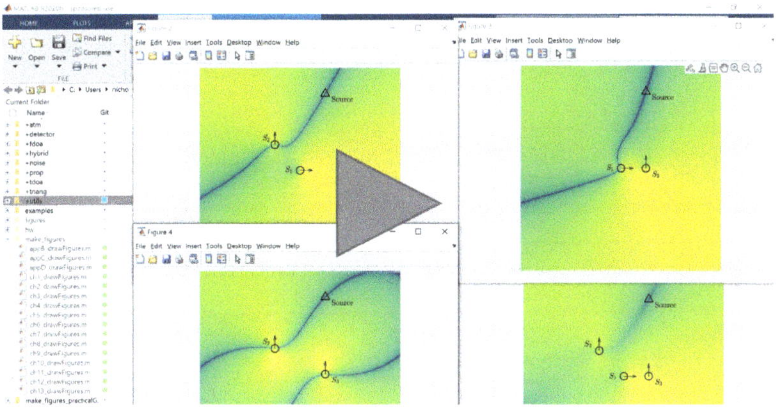

Figure 1.6 Video walkthrough of MATLAB® usage.

with tracking multiple targets, as well as Section 7.3 of [17] for a discussion of Kalman filters within a single sensor (radar) tracking context, and Section 7.3 and Chapter 9 of [18]. A formulation similar to that necessary for passive sensors can be found in [19], which contains a discussion of tracking in sensor coordinates (Chapter 7) and in target coordinates (Chapter 9).

1.3 ASSOCIATED SOFTWARE

This book is written with companion MATLAB® software to illustrate the concepts discussed, and to provide a functional starting point for readers to conduct their own research. The software is repeatedly referenced, and readers are strongly encouraged to run the examples and figure generation scripts as they read, to reinforce understanding.

1.3.1 MATLAB®

We use MATLAB® as our primary development environment, since it is widely employed as a modeling and simulation tool across the Department of Defense research enterprise, and is widely taught in engineering programs. We tested our code on MATLAB® R2020b, but do not expect significant errors with releases as old as R2015a.

Figure 1.6 is a quick introduction to the MATLAB® environment. The code was developed to be largely free of dependencies on special toolboxes, but users may want to consider the *Statistics & Machine Learning Toolbox*.[5]

5. There are workarounds in the +utils folder for some, but not all, functions referenced in the *Statistics & Machine Learning Toolbox*.

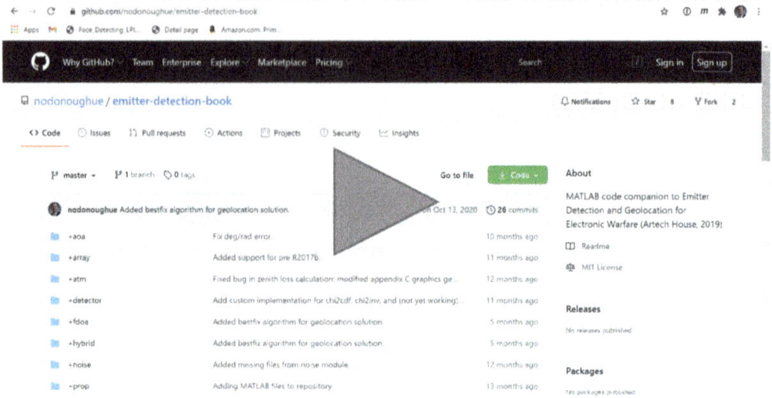

Figure 1.7 Video walkthrough of textbook software installation.

1.3.2 Textbook Software

The software for this text can be found in the git repository, which contains a combined package of source code for both this text and [2]. To use the code, simply download the repository as a .zip file and extract it to your computer, or clone the repository (if you are familiar with git). Then open MATLAB® in the extracted folder. For ease of use, you can run the command `setup`; this will add the folder to your MATLAB® path, so that you can use the tools every time you open MATLAB®.

```
>>setup
```

Figure 1.7 is a link to a video walkthrough of how to download, install, and run the provided source code from the git repository.

1.3.3 Python Software

At the time of publication, a port to Python was in process, but not yet complete. None of the examples, figures, or videos in this text reference the Python implementation, but users may access the latest version of the source code from the git repository.

References

[1] M. Esper, "Electromagnetic spectrum superiority strategy," Department of Defense, Tech. Rep., 2020.

[2] N. A. O'Donoughue, *Emitter Detection and Geolocation for Electronic Warfare*. Norwood, MA: Artech House, 2019.

[3] R. A. Poisel, *Electronic Warfare Receivers and Receiving Systems*. Norwood, MA: Artech House, 2015.

[4] D. Adamy, *EW 103: Tactical Battlefield Communications Electronic Warfare*. Norwood, MA: Artech House, 2008.

[5] R. Poisel, *Modern Communications Jamming: Principles and Techniques, 2nd Edition*. Norwood, MA: Artech House, 2011.

[6] H. L. Van Trees, *Detection, Estimation, and Linear Modulation: Part I of Detection, Estimation, and Modulation Theory*. Hoboken, NJ: Wiley, 2001.

[7] L. L. Scharf, *Statistical Signal Processing*. Reading, MA: Addison-Wesley, 1991.

[8] S. Boyd and L. Vandenberghe, *Convex Optimization*. Cambridge, UK: Cambridge University Press, 2004.

[9] J. Gorman and A. Hero, "Lower bounds for parametric estimation with constraints," *IEEE Transactions on Information Theory*, vol. 36, no. 6, pp. 1285–1301, 1990. DOI: 10.1109/18.59929.

[10] A. K. Jagannatham and B. D. Rao, "Cramer-rao lower bound for constrained complex parameters," *IEEE Signal Processing Letters*, vol. 11, no. 11, pp. 875–878, 2004. DOI: 10.1109/LSP.2004.836948.

[11] K. Ho and Y. Chan, "Geolocation of a known altitude object from TDOA and FDOA measurements," *IEEE Transactions on Aerospace and Electronic Systems*, vol. 33, no. 3, pp. 770–783, 1997. DOI: 10.1109/7.599239.

[12] U.S. Air Force, *RC-135U Combat Sent and RC-135V/W Rivet Joint*, Jun. 2012. [Online]. Available: https://www.flickr.com/photos/usairforce/7414270766/in/photolist-cib5oA.

[13] Lockheed Martin, *Lockheed Martin To Develop Cyber/Electronic Warfare Podded System for U.S. Army*, Jan. 2019. [Online]. Available: https://news.lockheedmartin.com/2019-01-19-Lockheed-Martin-to-Develop-Cyber-Electronic-Warfare-Podded-System-for-US-Army.pdf.

[14] U.S. Army, *Army modernizing electronic warfare capabilities*, Feb. 2019. [Online]. Available: https://www.army.mil/article/217353/army_modernizing_electronic_warfare_capabilities.pdf.

[15] C.-B. Chang and K.-P. Dunn, *Applied State Estimation and Association*. MIT press, 2016.

[16] K. Bell, T. Corwin, L. Stone, and R. Streit, *Bayesian Multiple Target Tracking, Second Edition*. Artech House, 2013.

[17] M. A. Richards, *Fundamentals of Radar Signal Processing*. New York, NY: McGraw-Hill Education, 2014.

[18] M. Skolnik, *Radar Handbook, 3rd Edition*. New York, NY: McGraw-Hill Education, 2008.

[19] M. Malanowski, *Signal Processing for Passive Bistatic Radar*. Norwood, MA: Artech House, 2019.

Chapter 2

Review of Emitter Geolocation

This chapter is a condensed review of estimation material from Chapter 6 of [1], and geolocation material from Chapters 9-13, with some additional discussion focused primarily on assisting with use of the provided MATLAB® code. Further review of estimation theory can be found in [2].

Before we begin, let us review some notation. The emitter that we are attempting to geolocate will be referred to throughout this text as simply the target, with coordinates $\mathbf{x}$. From the collection of sensors, we have a set of measurements ζ of an underlying set of parameters $\mathbf{z}$ (e.g., angle of arrival, time difference of arrival, or frequency difference of arrival) that relate to the unknown target position. This chapter discusses the relationship between $\mathbf{x}$, $\mathbf{z}$, and ζ.

To put it plainly, estimation theory describes the process by which we can estimate the unknown position $\mathbf{x}$ from the measurements ζ, and also describes the expected performance of that estimation process.

This chapter begins with a review of how AOA, TDOA, and FDOA measurements, and the information each provides with respect to target location and the likelihood function for each form of geolocation measurement, which forms the bases of geolocation. Then, we review various estimators, including *maximum likelihood (ML) estimation.* Finally, we briefly discuss common performance measures. In both cases, we outline how the provided MATLAB® code is used.

2.1 GEOLOCATION MEASUREMENTS AND THEIR LIKELIHOOD FUNCTIONS

If the measurement ζ is a random variable described by the position $\mathbf{x}$, then we can express the probability distribution of ζ as the *likelihood* that a given measurement ζ occurs as a function of the unknown position, written $\ell(\zeta|\mathbf{x})$. If we assume a Gaussian random variable, with an unknown mean given by the parameter $\mathbf{z}(\mathbf{x})$[1] and covariance

1. Where $\mathbf{z}(\mathbf{x})$ describes an ideal (noise-free) measurement.

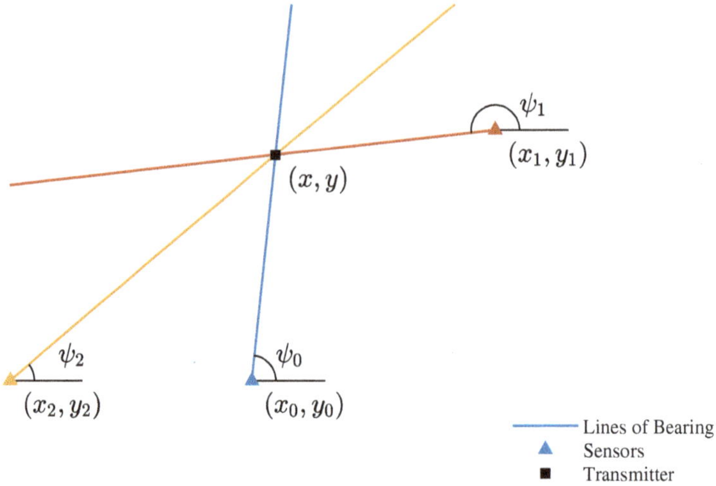

Figure 2.1 Recreation of Figure 10.2 from [1]. Example AOA geometry.

matrix $\mathbf{C}$, the likelihood function is

$$\ell(\zeta|\mathbf{x}) = \frac{1}{\pi^{-N/2}\left|\mathbf{C}_\zeta\right|^{-1/2}} e^{-\frac{1}{2}(\zeta-\mathbf{z}(\mathbf{x}))^H\mathbf{C}_\zeta^{-1}(\zeta-\mathbf{z}(\mathbf{x}))}. \tag{2.1}$$

Estimation theory states that we can equivalently replace the likelihood function $\ell(\zeta|\mathbf{x})$ with any similar *monotonic* function of $\ell(\zeta|\mathbf{x})$ [2]. In the case of Gaussian random variables, a popular substitution is the *log-likelihood* function, which (after dropping some constant terms) is:

$$\ell(\zeta|\mathbf{x}) = -\frac{1}{2}\left(\zeta - \mathbf{z}(\mathbf{x})\right)^H \mathbf{C}_\zeta^{-1}\left(\zeta - \mathbf{z}(\mathbf{x})\right). \tag{2.2}$$

Next, we review the log-likelihood for each of the three common geolocation techniques.

2.1.1 Likelihood Function for AOA Measurements

If the measurements ζ are angle of arrival measurements from multiple sensors, as shown in Figure 2.1, and there is no bias in the measurements (or if the bias is known and corrected), then the measurements can be expressed as a Gaussian random variable with mean value $\mathbf{z}(\mathbf{x}) = \boldsymbol{\psi}(\mathbf{x})$:[2.]

2. The target position $\mathbf{x} = (x, y)$ and the sensor position $\mathbf{x}_i = (x_i, y_i)$.

$$\psi_i(\mathbf{x}) = \tan^{-1}\left(\frac{y - y_i}{x - x_i}\right), \tag{2.3}$$

$$\boldsymbol{\psi}(\mathbf{x}) = \left[\psi_0, \dots, \psi_{N-1}\right]^T, \tag{2.4}$$

$$\zeta \sim \mathcal{N}\left(\boldsymbol{\psi}(\mathbf{x}), \mathbf{C}_{\boldsymbol{\psi}}\right), \tag{2.5}$$

where $\mathbf{C}_{\boldsymbol{\psi}}$ is the covariance matrix of the N AOA measurements. Under these assumptions, the log likelihood (ignoring scalars and terms that do not vary with $\boldsymbol{\psi}(\mathbf{x})$) is:

$$\ell(\zeta|\mathbf{x}) = -\frac{1}{2}\left(\zeta - \boldsymbol{\psi}(\mathbf{x})\right)^H \mathbf{C}_{\boldsymbol{\psi}}^{-1}\left(\zeta - \boldsymbol{\psi}(\mathbf{x})\right). \tag{2.6}$$

The parameters $\boldsymbol{\psi}(\mathbf{x})$ and likelihood function can be generated with the MATLAB® commands `measurement` and `log_likelihood`, from the `triang` package, respectively. For example:

```
psi = triang.measurement(x_sensor, x_source);
ell = triang.loglikelihood(x_sensor, psi+noise, cov, x_test);
```

where `x_sensor` is the set of coordinates for all sensors, `x_source` is the true target (source) position, `x_test` is a set of coordinates at which to evaluate the likelihood, `noise` is a noise vector to contaminate the AOA measurements, and `cov` is the covariance matrix used to generate the noise.

For more detail on AOA measurements, see Chapters 7-8 of [1], and for more discussion of geolocation based on AOA measurements, see Chapter 10 of [1].

2.1.1.1 2-D AOA Measurements

The formulation above is for a 2-D geometry in x and y coordinates. Extension to 3-D is straightforward, although it can be complex, as there are multiple conventions.

In this text, and the accompanying code, we use azimuth and elevation angles. While azimuth is defined above, the elevation angle (in the z coordinate) is given:

$$\phi_i = \tan^{-1}\left(\frac{z - z_i}{\sqrt{(x - x_i)^2 + (y - y_i)^2}}\right) \tag{2.7}$$

The provided code can compute both azimuth and elevation if 3-D coordinates are used.[3.] In this case, the measurement vector has length 2N, and is stacked:

$$\mathbf{z} = \left[\boldsymbol{\psi}^T(\mathbf{x}),\ \boldsymbol{\phi}^T(\mathbf{x})\right]^T \tag{2.8}$$

3. To prevent 2-D AOA calculations when using 3-D sensor positions, add a Boolean flag `do2DAOA=False` to the end of the function call to most the triang commands. The geometric solutions `triang.angle_bisector` and `triang.centroid` do not support 2-D AOA, so it is not necessary for those function calls.

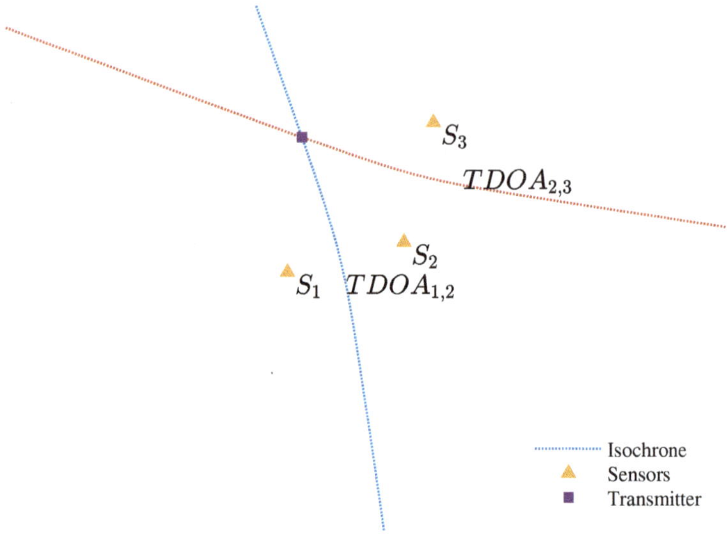

Figure 2.2 Recreation of Figure 11.1(b) from [1]. Example TDOA geometry.

and the covariance matrix should be similarly constructed:

$$\mathbf{C}_\psi = \begin{bmatrix} \mathbf{C}_\psi & \mathbf{0} \\ \mathbf{0} & \mathbf{C}_\phi \end{bmatrix} \tag{2.9}$$

where $\mathbf{C}_\psi$ is the covariance matrix of the azimuth angle measurements and $\mathbf{C}_\phi$ is the covariance matrix of the elevation angle measurements. The cross-terms $\mathbf{0}$ should be replaced with cross-covariance if the azimuth and elevation angle measurements are not independent.

Chapter 4 discusses 3-D coordinates in more detail.

2.1.2 Likelihood Function for TDOA Measurements

If the measurements ζ are time difference of arrival measurements from multiple sensor pairs, as shown in Figure 2.2, and there is no bias in the measurements (or if the bias is known and corrected), then the measurements can be expressed as a Gaussian random variable with mean value $\mathbf{z}(\mathbf{x}) = \mathbf{r}(\mathbf{x})$:[4].

4. Note $\mathbf{r}$ is actually the *range difference* rather than the *time difference*. This is achieved with a simple multiplication by the speed of light.

$$R_{i,j}(\mathbf{x}) = \|\mathbf{x} - \mathbf{x}_i\|^2 - \|\mathbf{x} - \mathbf{x}_j\|^2 \tag{2.10}$$

$$\mathbf{r}(\mathbf{x}) = \left[R_{1,N}(\mathbf{x}), \ldots, R_{N-1,N}(\mathbf{x})\right]^T, \tag{2.11}$$

$$\zeta \sim \mathcal{N}\left(\mathbf{r}(\mathbf{x}), \mathbf{C}_\mathbf{r}\right), \tag{2.12}$$

where $\mathbf{C}_\mathbf{r}$ is the covariance matrix of the $N-1$ TDOA measurements. In this formulation, the Nth sensor is used as a reference sensor, against which all other measurements are compared.[5.] The covariance matrix can thus be written:

$$\mathbf{C}_\mathbf{r} = c^2\left(\sigma_N^2 + \mathbf{diag}\left\{\sigma_1^2, \ldots, \sigma_{N-1}^2\right\}\right) \tag{2.13}$$

where σ_i^2 is the error variance of the ith time of arrival measurement, and c is the speed of light.

Under these assumptions, the log likelihood (ignoring scalars and terms that do not vary with $\mathbf{r}(\mathbf{x})$) is:

$$\ell(\zeta|\mathbf{x}) = -\frac{1}{2}\left(\zeta - \mathbf{r}(\mathbf{x})\right)^H \mathbf{C}_\mathbf{r}^{-1}\left(\zeta - \mathbf{r}(\mathbf{x})\right). \tag{2.14}$$

The measurements $\mathbf{r}(\mathbf{x})$ and likelihood function can be generated with the MATLAB® commands `measurement` and `log_likelihood`, from the `tdoa` package, respectively. For example:

```
r = tdoa.measurement(x_sensor, x_source);
ell = tdoa.loglikelihood(x_sensor, r+noise, cov, x_test);
```

where `x_sensor` is the set of coordinates for all sensors, `x_source` is the true target (source) position, `x_test` is a set of coordinates at which to evaluate the likelihood, `noise` is a noise vector to contaminate the TDOA measurements, and `cov` is the covariance matrix used to generate the noise.

Note that there is an optional input `ref_idx` to the TDOA functions provided that can be used to change which sensor is used as the reference. We will see in Chapter 3 how to also use that input to define a set of arbitrary sensor pairs.

For more detail on TDOA measurements and geolocation solutions, see Chapter 11 of [1].

2.1.3 Likelihood Function for FDOA Measurements

If the measurements ζ are frequency difference of arrival measurements from multiple sensor pairs, as shown in Figure 2.3, and there is no bias in the measurements (or if the bias is known and corrected), then the measurements can be expressed as a Gaussian random variable with mean value $\mathbf{z} = \dot{\mathbf{r}}(\mathbf{x})$:[6.]

5. Chapter 3 will discuss more general solutions with multiple reference sensors.
6. Note $\dot{\mathbf{r}}$ is actually the *range-rate difference* rather than the *frequency difference*. This is achieved with a simple multiplication by the speed of light and division by the carrier frequency.

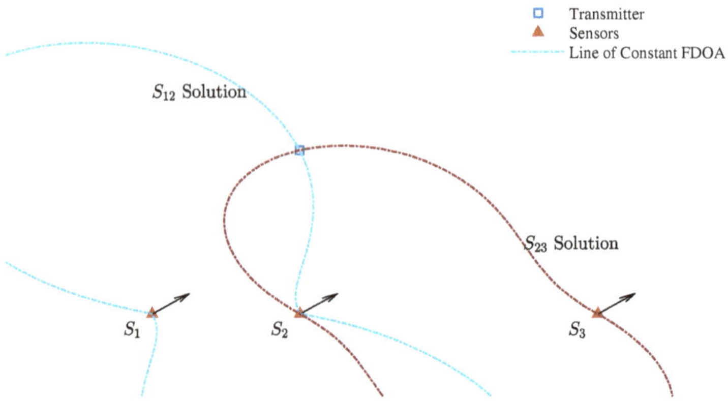

Figure 2.3 Recreation of Figure 12.1 from [1]. Example FDOA geometry.

$$\dot{R}_{i,j}(\mathbf{x}, \mathbf{v}) = \frac{(\mathbf{v} - \mathbf{v}_i)^T (\mathbf{x} - \mathbf{x}_i)}{\|\mathbf{x} - \mathbf{x}_i\|^2} - \frac{(\mathbf{v} - \mathbf{v}_j)^T (\mathbf{x} - \mathbf{x}_j)}{\|\mathbf{x} - \mathbf{x}_j\|^2} \tag{2.15}$$

$$\dot{\mathbf{r}}(\mathbf{x}, \mathbf{v}) = \left[\dot{R}_{1,N}(\mathbf{x}, \mathbf{v}), \dots, \dot{R}_{N-1,N}(\mathbf{x}, \mathbf{v})\right]^T, \tag{2.16}$$

$$\zeta \sim \mathcal{N}\left(\dot{\mathbf{r}}(\mathbf{x}), \mathbf{C}_{\dot{\mathbf{r}}}\right), \tag{2.17}$$

where $\mathbf{C}_{\dot{\mathbf{r}}}$ is the covariance matrix of the $N-1$ FDOA measurements. In this formulation, the Nth sensor is used as a reference sensor against which all other measurements are compared.[7] The covariance matrix can thus be written:

$$\mathbf{C}_{\dot{\mathbf{r}}} = \frac{c^2}{f_0^2}\left(\sigma_N^2 + \mathbf{diag}\left\{\sigma_1^2, \dots, \sigma_{N-1}^2\right\}\right) \tag{2.18}$$

where σ_i^2 is the error variance of the ith frequency measurement, c is the speed of light, and f_0 is the carrier frequency of the intercepted signal. Under these assumptions, the log likelihood (ignoring scalars and terms that do not vary with $\dot{\mathbf{r}}(\mathbf{x})$) is:

$$\ell(\zeta|\mathbf{x}) = -\frac{1}{2}\left(\zeta - \dot{\mathbf{r}}(\mathbf{x})\right)^H \mathbf{C}_{\dot{\mathbf{r}}}^{-1}\left(\zeta - \dot{\mathbf{r}}(\mathbf{x})\right). \tag{2.19}$$

The measurements $\dot{\mathbf{r}}(\mathbf{x})$ and likelihood function can be generated with the MATLAB® commands `measurement` and `log_likelihood`, from the `fdoa` package, respectively. For example:

```
r_dot = fdoa.measurement(x_sensor, v_sensor, x_source);
ell = fdoa.loglikelihood(x_sensor, v_sensor, r_dot+noise, cov, x_test);
```

7. Chapter 3 will discuss more general solutions with multiple reference sensors.

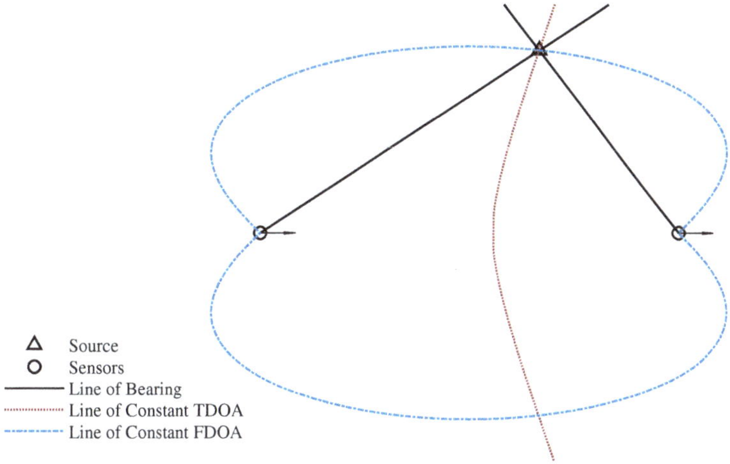

Figure 2.4 Recreation of Figure 13.1 from [1]. Example hybrid geolocation geometry.

where `x_sensor` is the set of coordinates for all sensors, `x_source` is the true target (source) position, `x_test` is a set of coordinates at which to evaluate the likelihood, `noise` is a noise vector to contaminate the FDOA measurements, and `cov` is the covariance matrix used to generate the noise.

For simplicity, we assumed $\mathbf{v} = \mathbf{0}$ in [1]. This is reflected in the code snippet here. We will discuss relaxing this assumption in Chapter 8.

Similar to the TDOA functions, there is an optional input `ref_idx` to the FDOA functions provided that can be used to change which sensor is used as the reference. We will see in Chapter 3 how to also use that input to define a set of arbitrary sensor pairs.

For more detail on FDOA measurements and geolocation solutions, see Chapter 12 of [1].

2.1.4 Likelihood Function for Hybrid Measurements

A hybrid geolocation system, in which a combination of AOA, TDOA, and FDOA measurements from some set of sensors (potentially disjoint, or colocated on a small number of platforms), has the potential to provide very robust geolocation with only a few measurements, as the errors associated with AOA, TDOA, and FDOA are often close to orthogonal, as shown in Figure 2.4 for an example geometry.

In this case, we simply stack the different measurement sets:

$$\mathbf{z} = \left[\boldsymbol{\psi}(\mathbf{x})^T, \mathbf{r}(\mathbf{x})^T, \dot{\mathbf{r}}(\mathbf{x}, \mathbf{v})^T\right]^T \tag{2.20}$$

The measurement vector ζ is similarly stacked, with covariance matrix:

$$\mathbf{C}_\zeta = \begin{bmatrix} \mathbf{C}_\psi & \mathbf{0}_{N,N-1} & \mathbf{0}_{N,N-1} \\ \mathbf{0}_{N-1,N} & \mathbf{C}_\mathbf{r} & \mathbf{C}_{\mathbf{r},\dot{\mathbf{r}}} \\ \mathbf{0}_{N-1,N} & \mathbf{C}_{\mathbf{r},\dot{\mathbf{r}}}^T & \mathbf{C}_{\dot{\mathbf{r}}} \end{bmatrix} \tag{2.21}$$

Here, we assume that the AOA measurements are independent of TDOA and FDOA (a reasonable assumption given how each is generated), but that TDOA and FDOA measurements are potentially correlated, as will be the case if they are generated from the same set of sensors.

Section 13.4.2 of [1] discusses joint estimation of time and frequency difference of arrival via cross-correlation of the received signal at two sensors. The result is that the error covariance for a single TDOA/FDOA joint estimate is:

$$\mathbf{C}_{r,\dot{r}} = \begin{bmatrix} \sigma_r^2 & \sigma_{r,\dot{r}} \\ \sigma_{r,\dot{r}}^* & \sigma_{\dot{r}}^2 \end{bmatrix} = \begin{bmatrix} 0 & ct_s & 0 \\ 0 & 0 & \frac{c}{2\pi f_0 t_s} \end{bmatrix} \mathbf{F}_{\phi,\kappa,\nu}^{-1} \begin{bmatrix} 0 & 0 \\ ct_s & 0 \\ 0 & \frac{c}{2\pi f_0 t_s} \end{bmatrix} \tag{2.22}$$

where $\mathbf{F}_{\phi,\kappa,\nu}$ is the Fisher Information Matrix of the phase, delay, and Doppler difference between the two received signals, derived in Section 13.4 of [1]. For simplicity, we do not repeat the complex structure of $\mathbf{F}_{\phi,\kappa,\nu}$ here, but it is included in the appendix. To compute the CRLB of joint time/frequency error for one or more sensors, readers can leverage the function `jointTimeFreqError` of the `hybrid` package in MATLAB®; it returns the CRLB for $\mathbf{C}_{r,\dot{r}}$.

From here, the log-likelihood takes a similar form as in the individual cases

$$\ell(\zeta|\mathbf{x},\mathbf{v}) = -\frac{1}{2}\left(\zeta - \mathbf{z}(\mathbf{x},\mathbf{v})\right)^H \mathbf{C}_\zeta^{-1}\left(\zeta - \mathbf{z}(\mathbf{x},\mathbf{v})\right). \tag{2.23}$$

The measurement and log likelihood can be generated programmatically from the `measurement` and `log_likelihood` commands of the `hybrid` package, in a similar manner as in the individual sensor type cases, although the function call is more complex. For more detail on geolocation from hybrid AOA/TDOA/FDOA measurements, see Chapter 13 of [1].

2.1.5 Generating Measurements and Likelihood Functions

As mentioned above, the `measurement` functions can be used to easily generate the AOA, TDOA, FDOA, or hybrid measurements for any arbitrary geometry. Those generated measurements can be used for idealized calculations, such as analysis of the inherent ambiguities in a given geometry, calculation of the likelihood function to understand the importance of sensor measurement error on those ambiguities, simulations such as Monte Carlo trials where random errors are added to ideal measurements in order to

Table 2.1
Sensor and Target Parameters for Example 2.1

Element	Position (x, y)	Velocity (x, y)	Sensor Type	Std. Dev.
Target	(3 km, 3 km)	0 m/s	–	–
Sensor 1	(4 km, 0 km)	0 m/s	AOA	3 deg
Sensor 2	(1 km, 0 km)	0 m/s	TDOA	100 ns
Sensor 3	(3 km, .5 km)	0 m/s	TDOA	100 ns
Sensor 4	(0 km, 1 km)	300 m/s @ -45 deg	FDOA	10 Hz
Sensor 5	(0 km, 2 km)	300 m/s @ -45 deg	FDOA	10 Hz

Std. Dev.: Measurement error standard deviation.

test geolocation algorithms and estimate their accuracy, or computation of predicted error performance (such as the statistical performance bounds we present later in this chapter). In this section, we present an example hybrid setup with a single AOA sensor, and individual pairs of TDOA and FDOA sensors, such that no single measurement type can geolocate the source, but where their collective measurements provide sufficient information to localize the target. We demonstrate the use of both the `measurement` and `log_likelihood` functions in this example.

The code to execute this example is provided in the `examples` folder.

Example 2.1 Relative Value of Hybrid AOA/TDOA/FDOA Measurements

Construct the hybrid sensor geometry in Table 2.1. Plot sensor and target positions. Generate the true sensor measurements for each sensor type, and overlay the ideal solutions. Then, compute and print both the sensor measurements, as well as the error variance for each type of data. For FDOA, assume a 1 GHz carrier frequency when converting from Doppler frequency (Hz) to range-rate (m/s) difference. Assemble the individual variances into a joint covariance matrix for all three measurements; assume independent measurements.

Then, plot the log likelihood for each individual sensor measurement type, as well as the joint likelihood for all measurements together. Comment on the structure of the ambiguities. For plotting purposes, use a color scale from $[-20, 0]$ dB.

Solution: The first step is to generate the geometry, which we can do with a few simple commands:

```
x_source = [3; 3]*1e3;          % Target (source) position

x_aoa = [4; 0]*1e3;             % AOA sensor position
x_tdoa = [1, 3; 0, .5]*1e3;     % TDOA sensor positions
x_fdoa = [0, 0; 1, 2]*1e3;      % FDOA sensor positions
```

```
v_fdoa = [1, 1; -1, -1]*sqrt(.5)*300; % FDOA sensor velocities
```

This, along with the AOA, TDOA, and FDOA solutions (lines of constant angle/time difference/frequency difference of arrival), are plotted in Figure 2.5. The solutions are generated by first calling the `measurement` function of each geolocation type (`triang`, `tdoa`, or `fdoa`), to generate the ideal measurement for each type, and then calling the `drawLob`, `drawIsochrone`, and `drawIsodop` functions of each type to generate the solutions, which are overlaid onto the geometry in Figure 2.5. Each of the outputs from those solution functions is a 2xN matrix with x-coordinates in the first row and y-coordinates in the second. In the TDOA and FDOA solution functions, the final two parameters are the number of points returned, and the maximum offset to consider (orthogonal to the vector connecting the two sensors).

```
psi = triang.measurement(x_aoa, x_source);
soln_aoa = triang.drawLob(x_aoa,psi,x_source,1.5);

r = tdoa.measurement(x_source, x_tdoa);
soln_tdoa = tdoa.drawIsochrone(x_tdoa(:,1), x_tdoa(:,2) ,...
            r(2)-r_tdoa(1), 1000, 5e3);

r_dot = fdoa.measurement(x_fdoa, v_fdoa, x_source);
soln_fdoa = fdoa.drawIsodop(x_fdoa(:,2),v_fdoa(:,2),x_fdoa(:,1), ...
          v_fdoa(:,1),r_dot,2000,15e3);
```

Note that while each solution in Figure 2.5 intersects with the target, they do so in different directions. This will be helpful, because ambiguities in one solution will be different than in the others, allowing for a unique solution when they are considered together.

The measurements were generated in the previous code snippet (`psi`, `r`, and `rr`, respectively). To compute their variances, we note that the AOA solution is provided in radians, so we need only convert the angle accuracy from Table 2.1 to radians and then square (to convert from error standard deviation to variance). TDOA error is similarly converted from time difference to range rate difference with a multiplication (by the speed of light), and then squared. FDOA error is multiplied by the speed of light as well, but then is divided by the carrier frequency. In the previous text, `cov_r` was asked for in terms of the sensor measurement pairs (with the function `loglikelihood` assuming that the final sensor was the common reference sensor), but it is now required in terms of the individual sensor measurements along with a specification for how the sensors are combined into time difference and frequency difference measurements; this concept is introduced in Chapter 3.

```
err_aoa = 3; % deg
cov_psi = (err_aoa*pi/180)^2; % rad^2
fprintf('AOA Measurement: %.2f deg\n',psi*180/pi);
fprintf('AOA Variance: %.2g rad^2\n',cov_psi);
```

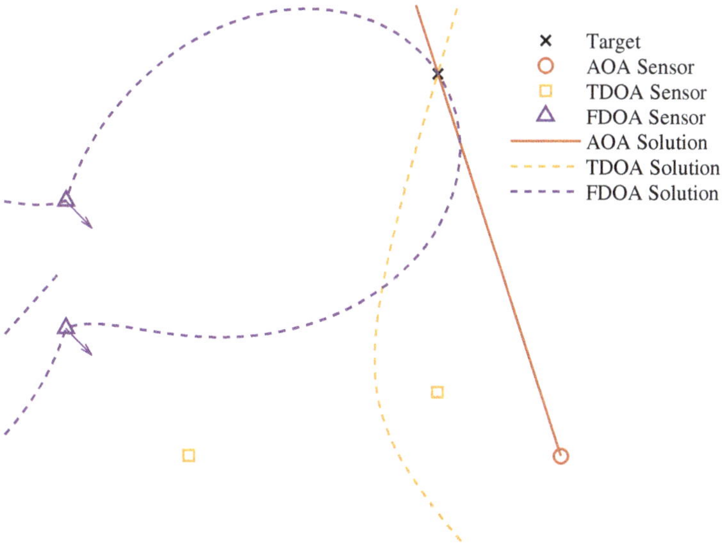

Figure 2.5 Geometry for Example 2.1, showing target location and sensor positions for each measurement type, as well as the partial solution (line of bearing, isochrone, and isodoppler contour for AOA, TDOA, and FDOA, respectively).

```
err_time = 1e-7; % 100 ns timing error
err_r = err_time * utils.constants.c;
n_tdoa = size(x_tdoa, 2);
cov_r = (err_r)^2 * eye(n_tdoa); % m^2
fprintf('TDOA Measurement: %.2f m\n',r);
fprintf('TDOA Variance: %.2g m^2\n',cov_r);

freq_err = 10; % Hz
f0 = 1e9; % Hz
rr_err = freq_err * utils.constants.c/f0; % (m/s)
n_fdoa = size(x_fdoa,2);
cov_rr = rr_err^2 * eye(n_fdoa); % (m/s)^2
fprintf('FDOA Measurement: %.2f m/s\n',rr);
fprintf('FDOA Variance: %.2g (m/s)^2\n',cov_rr);

% Hybrid Covariance Matrix
cov_z = blkdiag(cov_psi, cov_r, cov_psi);
```

which results in the output:

```
AOA Measurement: 108.43 deg
AOA Covariance: 0.0027 rad^2
```

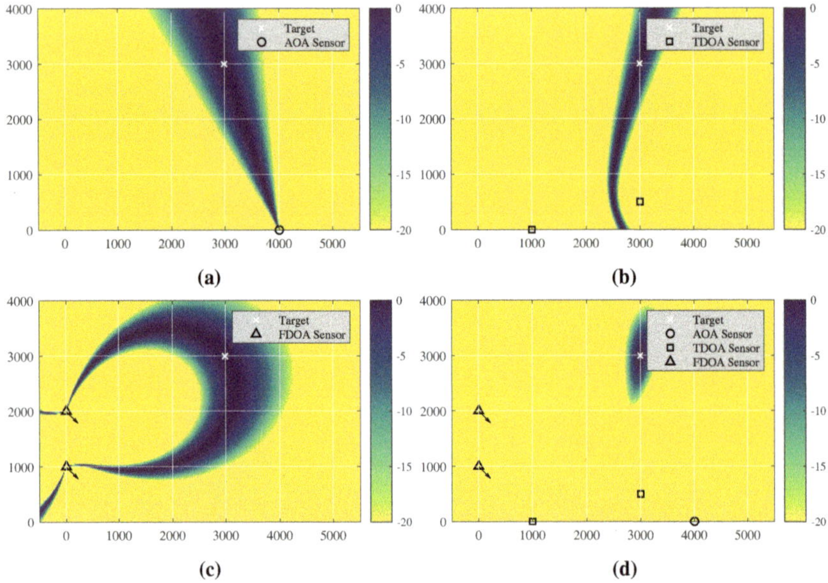

Figure 2.6 Plots of the component and joint likelihood functions for Example 2.1. (a) AOA only, (b) TDOA only, (c) FDOA only, and (d) all sensor measurements.

```
TDOA Measurement: 1105.55 m
TDOA Covariance: [898.76, 0.00
                                        0.00, 898.76] m^2

FDOA Measurement: -75.33 m/s
FDOA Covariance: [8.99, 0.00
                                        0.00, 8.99] (m/s)^2
```

The log likelihood functions are plotted in Figure 2.6. Figure 2.6(a) plots the AOA log likelihood, which shows a clear peak corresponding to the AOA solution in Figure 2.5, but with a thickness that arises from the error variance used to compute the likelihood. Similarly, Figure 2.6(b) plots the TDOA likelihood, which matches the TDOA solution in Figure 2.5, and Figure 2.6(c) plots the FDOA likelihood, which matches the FDOA solution in Figure 2.5. Figure 2.6(d) plots the joint likelihood of all three data sources together, exhibiting a clearly resolved peak near the true target location. □

Figure 2.7 is a video walkthrough of Example 2.1 and includes additional commentary.

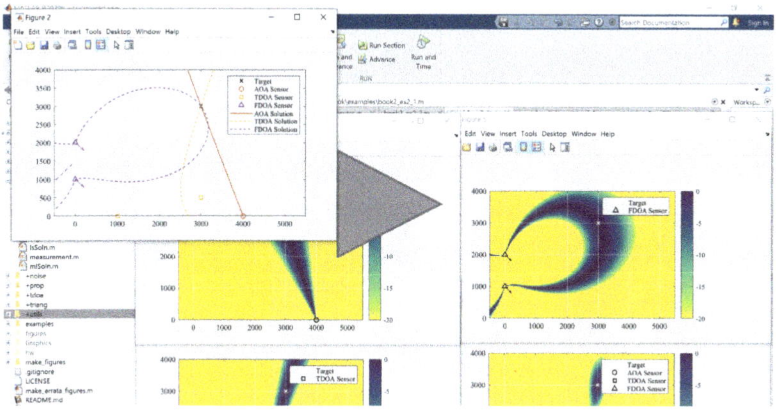

Figure 2.7 Video walkthrough of Example 2.1.

2.2 ESTIMATORS

In the provided code, a small set of classical estimators has been implemented: (1) a brute-force maximum likelihood approach, (2) a convex optimization approach, and (3) an iterative least-square approach. These are not meant to be exhaustive of the possible geolocation solvers, merely illustrative.

2.2.1 Maximum Likelihood Estimation

Maximum likelihood estimation is based around the simple concept of finding the parameter estimate $\widehat{\mathbf{x}}$ that maximizes the likelihood of the observed data $\boldsymbol{\zeta}$. Mathematically, this is written

$$\widehat{\mathbf{x}} = \arg\left[\max_{\mathbf{x}} \ell(\boldsymbol{\zeta}|\mathbf{x})\right] \tag{2.24}$$

This can be solved analytically by taking the gradient of the likelihood function with respect to $\mathbf{x}$ and finding all the points where it is zero (indicating a local minima or maxima). However, such analytic solutions are not typically available for geolocation, so we provided a numerical solution.[8].

The brute force solution to ML estimation involves creating a grid of test points, evaluating the likelihood of each point, and returning the coordinates of the maximum. The challenge of this solution is that if the grid is not sufficiently dense, it is possible to miss, or to generate a slightly inaccurate solution. However, if the grid is too dense, it can be computationally intractable to compute all of the likelihood calculations.

8. This topic is covered in detail in Chapter 6 of [2]; interested readers are directed there for a more thorough discussion.

In general, it is good practice to sample with grid density that is slightly finer than the expected geolocation accuracy, and that the search grid be no larger than is necessary to capture the possible set of solutions.

The MATLAB® software provided with this text includes a utility for brute force ML estimation, found in the `utils` package, in the function `ml_soln.m`. It can be accessed easily for AOA, TDOA, FDOA, or hybrid geolocation via the `ml_soln` function of the `triang`, `tdoa`, `fdoa`, and `hybrid` packages, respectively. See the `help` interface for usage instructions.

2.2.2 Maximum a Posterior Optimization

Maximum a Posterior (MAP) optimization is a generalization of ML estimation that seeks to take into consideration the joint distribution of both the observations and the parameters to be estimated, whereas ML estimation seeks to optimize the likelihood of the observations occurring. In mathematical terms, MAP is conducted via the optimization [2]:

$$\widehat{\mathbf{x}} = \arg\max_{\mathbf{x}} f_{\mathbf{x}}(\boldsymbol{\zeta}, \mathbf{x}) \tag{2.25}$$

Using Bayes formula, we can decompose the joint likelihood into two components, the conditional probability (from which we derived the likelihood function $\ell(\boldsymbol{\zeta}|\mathbf{x})$, and the (in the same manner as a joint PDF) with the equation:

$$f_{\mathbf{x}}(\boldsymbol{\zeta}, \mathbf{x}) = f_{\mathbf{x}}(\boldsymbol{\zeta}|\mathbf{x}) f_{\mathbf{x}}(\mathbf{x}) \tag{2.26}$$

If we plug this into (2.25), and take the logarithm (as we did for ML estimation), we arrive at:

$$\widehat{\mathbf{x}} \arg\max_{\mathbf{x}} \ell(\boldsymbol{\zeta}|\mathbf{x}) + \log\left(f_{\mathbf{x}}(\mathbf{x})\right) \tag{2.27}$$

Now, compare (2.27) with (2.24). The notable difference is the inclusion of the prior distribution on the solutions $f_{\mathbf{x}}(\mathbf{x})$. Thus, we can see that ML estimation is a special case of MAP, where the prior is uniform.

When considering cases with a prior distribution, we will use MAP instead of ML estimation as the basis of our discussion.

2.2.3 Convex Optimization

If the optimization problem is convex, meaning that both the *cost function* to be optimized and any constraints on the solution space are both convex, then it can be shown that a local optimum will also be global, and this allows for the use of iterative searches that are much more computationally tractable than brute force grid searches [3]. In fact, Boyd argues that this is an incredibly powerful tool for solving complex optimization problems. If they can be shown to be convex, or can be reformulated in such a way as to be convex, then solving them analytically is not necessary, since the numerical tools are efficient and guaranteed to converge.

Geolocation of emitters is not strictly convex, but it can be reasonably approximated as such in many circumstances, particularly if the search grid is localized near the target area. Thus, we included convex optimization in our MATLAB® code, using gradient descent with a backtracking line search to compute the iterative solutions. The code for this algorithm is found in the `gdSoln` and `backtrackingLineSearch` functions of the `utils` package. For convenience, they can be accessed via the `gdSoln` functions in the `triang`, `tdoa`, `fdoa`, and `hybrid` packages for AOA, TDOA, FDOA, and hybrid geolocation, respectively.

2.2.4 Least Square Estimators

In least squares estimation, the goal is to minimize the sum of the squared error term.

$$\epsilon(\mathbf{x}) = [\zeta - \mathbf{z}(\mathbf{x})]^H [\zeta - \mathbf{z}(\mathbf{x})] \tag{2.28}$$

The least squares estimator minimizes this error

$$\widehat{\mathbf{x}} = \arg\left[\min_{\mathbf{x}} \epsilon(\mathbf{x})\right] \tag{2.29}$$

If the function $\mathbf{z}(\mathbf{x})$ is linear, then we can rewrite it $\mathbf{Zx}$, and the solution to the optimization problem is:

$$\widehat{\mathbf{x}} = \left(\mathbf{G}^H\mathbf{G}\right)^{-1}\mathbf{G}^H\zeta \tag{2.30}$$

Our problems, however, are not linear. In [1], we introduced an iterative least square solution where the residual $\mathbf{y}(\mathbf{x}) = \zeta - \mathbf{z}(\mathbf{x})$ is approximated with a first-order Taylor series expansion about the point $\mathbf{x}$, and the least square solution in (2.30) is used to compute a step value $\Delta\mathbf{x}$ closer to the true target position. The step value is computed

$$\mathbf{y}\left(\mathbf{x}^{(i)}\right) = \zeta - \mathbf{z}\left(\mathbf{x}^{(i)} + \Delta\mathbf{x}^{(i)}\right) \tag{2.31}$$

$$\Delta\mathbf{x}^{(i)} = \left(\mathbf{J}\left(\mathbf{x}^{(i)}\right)\mathbf{C}_\zeta^{-1}\mathbf{J}^T\left(\mathbf{x}^{(i)}\right)\right)^{-1}\mathbf{J}\left(\mathbf{x}^{(i)}\right)^{-1}\mathbf{y}\left(\mathbf{x}^{(i)}\right). \tag{2.32}$$

This is used to update the position estimate:

$$\mathbf{x}^{(i+1)} = \mathbf{x}^{(i)} + \Delta\mathbf{x}^{(i)}, \tag{2.33}$$

which is iterated until $\Delta\mathbf{x}^{(i)}$ is below a desired resolution.[9.] Note that we include the covariance matrix for the measurements $\mathbf{C}_\zeta$ in our step calculation. This is a deviation from the strict formulation of iterative least squares, and is referred to as *weighted* least squares. The purpose of including the covariance matrix is to allow for greater

9. For stability, the provided code also checks for divergence, indicative of a poorly defined problem, and accepts a cap on the number of iterations.

weight to be placed on those measurements that are more accurately known, while those with greater variance are less heavily weighted. It also (in the case of a nondiagonal covariance matrix) allows for consideration of dependence between measurements.

The code that executes our numerical LS solution is found in the `lsSoln` file of the `utils` package, and it can be accessed for AOA, TDOA, FDOA, and hybrid scenarios via the `lsSoln` method of the `triang`, `tdoa`, `fdoa`, and `hybrid` packages, respectively.

2.2.5 Using Estimators for Geolocation

The provided estimators (`mlSoln`, `gdSoln`, and `lsSoln`) can be used to quickly assemble a geolocation estimate. To demonstrate their utility, this section provides and discusses an example geolocation problem.

Example 2.2 Target Location Estimation with Hybrid Measurements

For the scenario outlined in Example 2.1, generate a random set of errors for each measurement, from a real Gaussian distribution with variances as in Table 2.1. Attempt maximum likelihood, convex, and least squares optimization to estimate the source position. For gradient descent and least squares, use a seed position of (2.5 km, 2.5 km) as the initial estimate. For the ML solution, use a grid spanning (0, 0) in one corner, and (5 km, 5 km) in the opposite, with 10 m $\times$ 10 m grid resolution. Use all of the measurements to generate a solution. Plot the geometry, as well as the ML solution, and the first 100 iterations of both the convex and least square estimators.

Solution: The first thing that we do is generate a noisy measurement. The covariance terms `cov_psi`, `cov_r`, and `cov_rr` are as defined in Example 2.1. In addition to the sensor measurement covariance matrix `cov_z`, we need a covariance matrix for the sensor pairs, which we'll call `cov_z_out`, and will generate with the command `resampleCovMtx`, using the parameter `1` to indicate that the first sensor is the common reference. This is, once again, a result of the new sensor selection interface discussed at length in Chapter 3.

```
% Generate covariance matrix for sensor pairs (for TDOA/FDOA)
cov_r_out = utils.resampleCovMtx(cov_r,1); % TDOA sensor pairs
cov_rr_out = utils.resampleCovMtx(cov_rr,1); % FDOA sensor pairs
cov_z_out = blkdiag(cov_psi, cov_r_out, cov_rr_out); % Combined msmts
```

Using those values, we generate a hybrid measurement, and use the `randn` to generate Gaussian noise. The Cholesky decomposition is used to find the upper-right matrix $\mathbf{R}$ such that $\mathbf{C} = \mathbf{R}^H\mathbf{R}$, which allows us to generate correlated noise from the uncorrelated outputs of `randn` [4].[10.]

10. Alternatively, if you have the *Statistics & Machine Learning Toolbox*, you can use the covariance matrix and the function `mvnrnd` to generate correlated Gaussian noise.

```
% Hybrid measurement
z = hybrid.measurement(x_aoa, x_tdoa, x_fdoa, v_fdoa, x_source);

% Generate Random Noise
L = chol(cov_z, 'lower');
noise = L * randn(size(L,1), 1);

% Noisy Measurements
zeta = z + noise;
```

Next, set up the search parameters. The ML solver requires a regular grid, specified by its center, size, and spacing. The GD and LS solution both require an initial guess, desired resolution, and (optionally) the maximum number of iterations. The GD solution also accepts `alpha` and `beta` parameters for the backtracking line search that is repeated at every iteration (see the function `backTrackingLineSearch` in the `utils` package), but we can pass in empty brackets to use the default settings. The last two inputs for GS and LS are the `force_full_calc` and `plot_progress` Boolean flags. The former prevents the search algorithm from terminating before it reaches the desired number of iterations, and the latter controls whether the algorithm generates interim plots as it progresses.

```
%% ML Search Parameters
x_ctr = [2.5; 2.5]*1e3;
grid_size = [5e3; 5e3];
grid_res = 25;  % meters, grid resolution

%% GD and LS Search Parameters
x_init = [1; 1]*1e3;
epsilon = grid_res;
max_num_iterations = 100;
force_full_calc = true;
plot_progress = false;
```

Then, it's simply a matter of calling the three solvers. For the ML solver, the only output is the estimated target position, whereas the GD and LS solvers both provide the final estimate as the first output argument, and the matrix of intermediate estimates as the second output argument. The latter is useful for illustrating the way in which these iterative solvers arrive at their solutions.

```
%% ML Soln
x_ml = hybrid.mlSoln(x_aoa, x_tdoa, x_fdoa, v_fdoa, zeta, cov_z, x_ctr,
    grid_size, grid_res);

%% GD Soln
[x_gd, x_gd_full] = hybrid.gdSoln(x_aoa, x_tdoa, x_fdoa, v_fdoa, zeta, cov_z
    , x_init, [], [], epsilon, max_num_iterations, true, false);
```

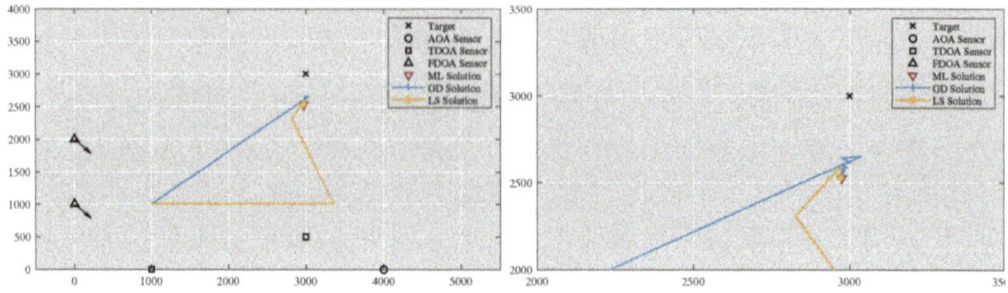

(a) (b)

Figure 2.8 Plot of the maximum likelihood, gradient descent, and least squares solutions for Example 2.2. (a) Full scenario, and (b) close look at target area.

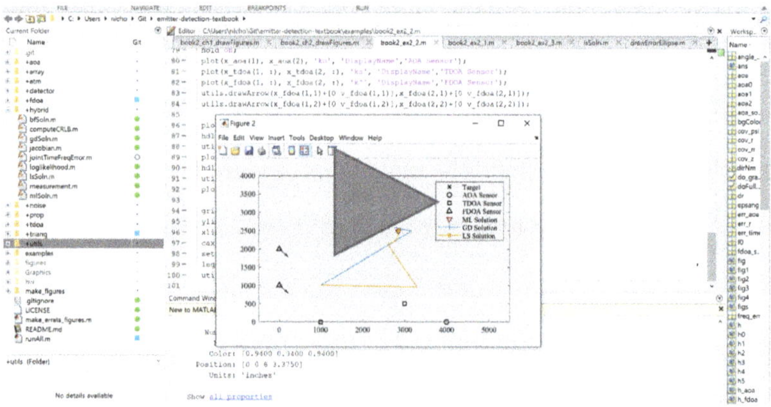

Figure 2.9 Video walkthrough of Example 2.2.

```
%% LS Soln
[x_ls, x_ls_full] = hybrid.lsSoln(x_aoa, x_tdoa, x_fdoa, v_fdoa, zeta, cov_z
    , x_init, epsilon, max_num_iterations, true, false);
```

Figure 2.8 shows the example geometry, with the estimates. For the GD and LS solvers, the iterative estimates are shown as lines, and the final solution is marked. Figure 2.8(a) shows the full scenario, while Figure 2.8(b) shows a closer look at the target area. You can see that all three solvers yields a similar solution, that is roughly 500 m from the true target position. This error is a result of the noisy measurements. □

Figure 2.9 contains a video walkthrough of Example 2.2 with additional commentary.

2.3 OTHER GEOLOCATION ALGORITHMS

As discussed in Chapters 11 and 12 of [1], a few custom algorithms for AOA and for TDOA geolocation were implemented in addition to the classical algorithms we've described. For AOA, we implemented a few geometric solutions, namely *centroid* and *angle bisection* geolocation, available as the `centroid` and `angle_bisector` methods of the `triang` package. For TDOA, we implemented a closed-form TDOA solution by Chang and Ho [5], available as the `chanHoSoln` method of the `tdoa` package. Other extensions and variations on this algorithm have been published by Prof. Ho's group [6]–[10], and similar solutions abound in the literature.

2.4 PERFORMANCE MEASURES

For any of the estimators described so far, the estimate $\widehat{\mathbf{x}}$ is properly viewed as a random vector, since it is the result of a deterministic function (the estimator) applied to a random vector (the observation $\boldsymbol{\zeta}$). Performance can be described by analyzing the distribution of the error term

$$\boldsymbol{\epsilon} = \widehat{\mathbf{x}} - \mathbf{x} \tag{2.34}$$

Most commonly, we analyze the first and second moments of the distribution of $\boldsymbol{\epsilon}$, referred to as the bias and error covariance, respectively

$$\mathbf{b} = E\left\{\boldsymbol{\epsilon}\right\} = E\left\{\widehat{\mathbf{x}} - \mathbf{x}\right\} \tag{2.35}$$

$$\mathbf{C}_\epsilon = E\left\{(\boldsymbol{\epsilon} - \mathbf{b})(\boldsymbol{\epsilon} - \mathbf{b})^H\right\} \tag{2.36}$$

For an *unbiased* estimator, $\mathbf{b} = \mathbf{0}$, and we can rewrite the error covariance

$$\mathbf{C}_\epsilon = E\left\{(\widehat{\mathbf{x}} - \mathbf{x})(\widehat{\mathbf{x}} - \mathbf{x})^H\right\} \tag{2.37}$$

Most performance measures rely on either the bias or the error covariance.

2.4.1 Root Mean Squared Error (RMSE)

A common error calculation in evaluating estimators is the RMSE. It is commonly defined for a set of sample errors, as in a Monte Carlo simulation or experimental test results, but can be equivalently predicted with the trace of the error covariance $\mathbf{C}_\epsilon$:

$$\text{RMSE} = \sqrt{\frac{1}{N}\sum_{i=1}^{N}\left\|\widehat{\mathbf{x}}_i - \mathbf{x}\right\|^2} = \sqrt{\text{Tr}\left\{\mathbf{C}_\epsilon\right\}} \tag{2.38}$$

where $\widehat{\mathbf{x}}_i$ is the ith estimate of $\mathbf{x}$.

2.4.2 Error Ellipse

For any two dimensions, the error ellipse can be used to trace the distribution of estimates about the bias point. It is computed simply from the ellipse equation

$$\left(\frac{x - b_x}{\sigma_x}\right)^2 + \left(\frac{y - b_y}{\sigma_y}\right)^2 = \gamma \tag{2.39}$$

where b_x and b_y are the x and y components of the bias vector $\mathbf{b}$, σ_x and σ_y are the square roots of the x and y variance terms from $\mathbf{C}_\epsilon$, respectively, and γ is a scale parameter used to determine the area of the ellipse. The selection of γ controls what percentage of estimates fall within the ellipse; thereby defining its confidence interval, while the variance terms define its shape and the bias terms its centroid.

This formulation only applies if the x and y components of the position error are independent, which is almost never the case. Thus, the computation of the error ellipse must first find the semi-major and semi-minor axes of the error ellipse, then define the ellipse in those terms, and then rotate to align with the x and y axes. This math is described in Section 9.2 of [1]. The provided MATLAB® code implicitly handles the necessary rotations from the provided covariance matrix. To generate an error ellipse, simply call the `drawErrorEllipse` method of the `utils` package, with the center of the ellipse (the estimated target position), the covariance matrix, the number of points to compute (to trace the ellipse), and the desired confidence interval (a larger ellipse is more likely to contain the true position). Defined confidence intervals are 1 (1 standard deviation – roughly 67%), 50 (50%), 90 (90%), and 95 (95%).

```
>> x_est = [2;1];
>> cov= [1 3; 2 4];
>> num_pts = 101;
>> conf_interval = 50;
>> [x_ellipse, y_ellipse] = utils.drawErrorEllipse(x_est, cov, num_pts,
     conf_interval);
```

In three dimensions, the error ellipse is replaced by an error ellipsoid. Typically 3-D errors are represented by their projections onto the various 2-D planes (x-y, y-z, and x-z).

2.4.3 Circular Error Probable (CEP)

CEP is defined simply as the radius of a circle—centered on the emitter location—such that a given percentage of estimates fall within that circle. The most commonly used metrics are CEP_{50}—wherein 50% of the estimates fall within the circle—and CEP_{90}—wherein 90% of the estimates fall within the circle.

The definition of the CEP is rather straightforward, but its computation is not; papers studying the CEP go back to the 1950s [11], [12]. The most popular approximation for CEP_{50} is [13], [14]

$$\text{CEP}_{50} = \begin{cases} .59\left(\sigma_s + \sigma_l\right) & \frac{\sigma_s}{\sigma_l} \geq .5 \\ \sigma_l\left(.67 + .8\frac{\sigma_s^2}{\sigma_l^2}\right) & \frac{\sigma_s}{\sigma_l} < .5 \end{cases} \tag{2.40}$$

where the error terms σ_s and σ_l are, respectively, the smaller and larger square roots of the eigenvalues of the covariance matrix $\mathbf{C}_\epsilon$. The first approximation is valid when the error ellipse has low eccentricity (the two error terms are approximately equal), and the latter is used when one error term dominates.

Just as with the error ellipse, CEP is typically computed in two dimensions and can be done for (x, z) or (y, z) in addition to the (x, y) variant shown here. However, since CEP is scalar, there is also the opportunity to include all three error terms; the result is *spherical error probable* (SEP). A simple approximation to SEP, for 50% and 90%, respectively, is given [15]

$$SEP_{50} \approx .51\left(\sigma_x + \sigma_y + \sigma_z\right) \tag{2.41}$$

$$SEP_{90} \approx .833\left(\sigma_x + \sigma_y + \sigma_z\right) \tag{2.42}$$

These approximations also assume that the $\hat{\mathbf{x}}$ is unbiased. Additional approximations were reported in [16], derived through curve fitting, and reported to have accuracy of roughly 2% across a broad range of error variance ratios between the spatial dimensions.

In the provided MATLAB® code, there is a pair of functions provided for CEP calculation, in the `utils` package, the function `computeCEP50` will compute the CEP_{50} from a covariance matrix. Similarly, the function `drawCEP50` will generate a circle with radius matching the CEP_{50} from a center point and a covariance matrix.

2.4.4 Statistical Bounds

If a system exists, it can be tested or simulated to obtain an estimate of the error covariance matrix ($\mathbf{C}$) and bias ($\mathbf{b}$), and any of the measures above can be computed. The downside is that is typically a slow and costly process and does not easily extend to situations that were not tested. Statistical bounds, however, can typically be solved analytically or numerically, allowing for insight and intuition into the behavior of the underlying algorithm and providing performance guarantees across a greater range of scenarios.

2.4.4.1 Cramér-Rao Lower Bound

One popular, and very useful, performance measure is the Cramér-Rao Lower Bound (CRLB) [2], [17]–[19]. This is a lower bound on the error covariance $\mathbf{C}$ for an unbiased estimator:

$$\mathbf{C}\left(\hat{\mathbf{x}}\right) \geq CRLB \tag{2.43}$$

The reason the CRLB is very useful is that it is often achievable and can thus be used to guarantee that a given unbiased estimator is *efficient* (meaning that there is no alternative estimator with a lower error covariance for a given input SNR). Since there is no unbiased estimator that can violate (2.43), any unbiased estimator that achieves it is optimal in that sense.

The CRLB is the inverse of the Fisher information matrix [2], which can be expressed (for Gaussian measurements) in terms of the measurement covariance matrix and the Jacobian matrix (defined as the gradient of the log-likelihood with respect to all of the unknown parameters):

$$\mathbf{J_z}(\mathbf{x}) \triangleq [\nabla_{\mathbf{x}} z_0(\mathbf{x}),\ \nabla_{\mathbf{x}} z_1(\mathbf{x}), \dots, \nabla_{\mathbf{x}} z_{N-1}(\mathbf{x})] \tag{2.44}$$

$$\mathbf{F_z}(\mathbf{x}) = \mathbf{J_z}(\mathbf{x})\mathbf{C}_\zeta^{-1}\mathbf{J}_\mathbf{z}^T(\mathbf{x}) \tag{2.45}$$

$$\mathbf{C}_\epsilon(\mathbf{x}) \geq \mathbf{F}_\mathbf{z}^{-1}(\mathbf{x}) \tag{2.46}$$

where $\nabla_{\mathbf{x}} z_n(\mathbf{x})$ is the gradient of the equation for the nth ideal (noise-free) measurement with respect to each dimension of the target position.

For a more complete definition, including treatment of nuisance parameters, and alternate formulations of the Fisher information matrix, interested readers are referred to Sections 6.4 and 6.5 of [2] or Section 6.3 of [20]. We will discuss this in greater detail, when we introduce constraints to the CRLB, in Chapter 6.

The CRLB is not the only statistical bound that is relevant to geolocation performance, but it is the most widely cited; this is largely due to its ease of derivation and computation, as well as its achievability; an unbiased maximum likelihood estimator will asymptotically achieve the Cramér-Rao lower bound as SNR tends to infinity.

One of the principal complaints about the CRLB is that it is overly optimistic in low SNR (large error) situations; it is also limited in that it cannot apply to *biased* estimators. In the following sections, we will briefly introduce some popular alternatives that are relevant to geolocation estimation. These bounds are generally grouped into *small error* (high SNR) and *large error* (low SNR) regimes, with the small error bounds typically relying on information theory while the large error bounds rely on sampling to estimate performance.

2.4.4.2 Barankin Bound

The Barankin bound [19], [21], [22], considers not just the behavior of the log-likelihood functions near the true parameter $\mathbf{x}$, as the CRLB does, but also the behavior at distant locations. This has the effect of more accurately capturing the likelihood and impact of larger errors or ambiguities [19]. The Barankin bound achieves this by selecting a finite number of test points $\mathbf{x}_i,\ i = 1, \dots, M$ and computing a measure of the error bound for these test points. The form of this bound is given:

$$C_\zeta \geq \mathbf{\Phi}(\mathbf{x})\left(\mathbf{B}(\mathbf{x}) - \mathbf{1}\mathbf{1}^T\right)^{-1}\mathbf{\Phi}^T(\mathbf{x}) \tag{2.47}$$

$$\mathbf{\Phi}(\mathbf{x}) = [\mathbf{x}_1 - \mathbf{x}, \mathbf{x}_2 - \mathbf{x}, \ldots, \mathbf{x}_M - \mathbf{x}] \tag{2.48}$$

$$[\mathbf{B}(\mathbf{x})]_{i,j} = E_{\mathbf{x}}\left\{\frac{\ell(\zeta|\mathbf{x}_i)}{\ell(\zeta|\mathbf{x})}\frac{\ell(\zeta|\mathbf{x}_j)}{\ell(\zeta|\mathbf{x})}\right\}, \; 1 \leq i,j \leq M \tag{2.49}$$

In other words, $\mathbf{\Phi}$ represents the distance from each test point to the point where the bound is evaluated, and the i, jth element $\mathbf{B}$ is the difference in likelihood between the ith and jth test points and the point where it is being evaluated, a measure of how likely those errors are to occur.

In order to form a very tight bound, however, many different families of test points must be evaluated, and optimized across (the largest error bound is kept). This can be cumbersome and somewhat costly to achieve confidence in a tight error bound.

2.4.4.3 Bhattacharyya Bound

A third bound that is also relevant is the Bhattacharyya bound [9], [23], [24], which is more complex but can provide a lower bound on estimation performance for *biased* estimators. The mth order Bhattacharyya inequality is given:

$$\mathbf{C}_\zeta \geq \mathbf{b}(\mathbf{x})\mathbf{b}^T(\mathbf{x}) + \boldsymbol{\alpha}_m\mathbf{K}_m^{-1}\boldsymbol{\alpha}_m^T \tag{2.50}$$

where $\boldsymbol{\alpha}_m$ is a *translation matrix* and $\mathbf{K}_m$ is the mth order *Bhatacharyya* matrix.[11]. If $\mathbf{K}_m$ is nonsingular, then it can be shown that the bound is the same as the CRLB (if the bias term is zero), and that as the order grows, the bound increases monotonically, meaning that a higher order m of Bhattacharyya bound will be *at least as tight* as any lower-order bound, and potentially more closely reflect what is achievable.

2.4.4.4 Hammersley-Chapman-Robbins Inequality

The Hammersley-Chapman-Robbins inequality [25], [26] is a less computationally intensive form of the Bhattacharyya. The *Hammersley-Chapman-Robbins information matrix* for a series of test points $\mathbf{x}_i,\; i = 1, \ldots, M$ is given [24]:

$$[\widetilde{\mathbf{J}}(\mathbf{x})]_{i,j} = E_{\mathbf{x}}\left\{\frac{\ell(\zeta|\mathbf{x}_j)}{\ell(\zeta|\mathbf{x})}\right\} \tag{2.51}$$

To compute the HCR bound, simply replace the matrix $\mathbf{B}(\mathbf{x})$ in (2.47) with $\widetilde{\mathbf{J}}(\mathbf{x})$. Computationally, this is a straightforward bound to compute, but it still requires testing

11. Their specific structure is not critical to our discussion, but they are built on a series of mth order gradients with respect to the parameter vector. $\mathbf{K}_1$ is the same as the Fisher information matrix, but as m grows, the order of the partial derivatives increases.

multiple sets of test points $\mathbf{x}_i$ and returning the largest error bound from across those sets.

2.4.4.5 Hybrid Bounds

In order to leverage the accuracy of small error bounds and large error bounds, in their respective regimes, many hybrid bounds have been derived. One such bound is the *hybrid Bhattacharyya-Barankin* (HBB) bound [24], [27]. The (m, r)th order HBB is essentially the sum of the mth order Bhatacharyya bound, and something like an rth order HCR bound (which is itself a simplification of the Barankin bound). Abel applied this to DOA estimation [24], and it has recently been used for hybrid AOA/TDOA geolocation, as well [9], [10]. Van Trees and Bell showed that the HBB is a generalization of many other hybrid bounds, in that they can be formulated as special cases of the HBB [27].

2.4.4.6 Ziv-Zakai Bound

The Ziv-Zakai Bound (ZZB) frames the estimation problem as a binary hypothesis test between the true parameter value and a slightly perturbed one, and then computes the smallest perturbation for which the test can successfully discriminate[28]–[32]; this is sometimes called a *Bayesian* bound [19], since a prior distribution on the parameter can be used to weight the computed errors. The advantage of the ZZB is that it is much tighter at moderate and low SNR levels than the CRLB, and that it can apply to biased estimators. The disadvantage is that evaluation of the ZZB almost universally involves a numerical integration, and so it is sometimes very costly to compute.

2.4.5 Statistical Bounds for Common Geolocation Data

In the following sections, we review the calculation of the CRLB, specifically the form of the Jacobian matrix and the error covariance matrix, for each of the four geolocation phenomenologies. In each case, the assumption is that the sensor measurements are Gaussian and, thus, the CRLB takes the form in (2.44-2.46).

2.4.5.1 CRLB for AOA Measurements

For AOA, where $\mathbf{z}(\mathbf{x}) = \boldsymbol{\psi}(\mathbf{x})$ as outlined in (2.3), the Jacobian is given in (10.10) of [1] and is expanded here for three dimensions, and for both azimuth (ψ) and elevation (ϕ)

angles:

$$\mathbf{J_z}(\mathbf{x}) = [\mathbf{J}_\psi(\mathbf{x})\mathbf{J}_\phi(\mathbf{x})] \tag{2.52}$$

$$= [\nabla_\mathbf{x}\psi_0(\mathbf{x}), \ldots, \nabla_\mathbf{x}\psi_{N-1}(\mathbf{x}), \nabla_\mathbf{x}\phi_0(\mathbf{x}), \ldots, \nabla_\mathbf{x}\phi_{N-1}(\mathbf{x})] \tag{2.53}$$

$$\nabla_\mathbf{x}\psi_n(\mathbf{x}) = \frac{1}{\|\mathbf{x} - \mathbf{x}_n\|^2} \begin{bmatrix} -(y - y_n) \\ (x - x_n) \\ 0 \end{bmatrix} \tag{2.54}$$

$$\nabla_\mathbf{x}\phi_n(\mathbf{x}) = \frac{1}{\|\mathbf{x} - \mathbf{x}_n\|^2} \begin{bmatrix} \frac{-(x-x_n)(z-z_n)}{R_g(\mathbf{x})} \\ \frac{-(y-y_n)(z-z_n)}{R_g(\mathbf{x})} \\ R_g(\mathbf{x}) \end{bmatrix} \tag{2.55}$$

$$R_g(\mathbf{x}) \triangleq \sqrt{(x - x_n)^2 + (y - y_n)^2} \tag{2.56}$$

For AOA measurements $\mathbf{C}_\zeta$ is typically diagonal,

$$\mathbf{C}_\zeta = \text{diag}\left\{\sigma_0^2, \ldots, \sigma_{L-1}^2\right\} \tag{2.57}$$

where σ_l^2 is the error of the lth AOA sensor measurement (in radians squared).

2.4.5.2 CRLB for TDOA Measurements

For TDOA, where $\mathbf{z}(\mathbf{x}) = \mathbf{r}(\mathbf{x})$ as outlined in (2.11), the Jacobian is given in (11.10) of [1]:

$$\mathbf{J_z}(\mathbf{x}) = \mathbf{J_r}(\mathbf{x}) = [\nabla_\mathbf{x}R_{1,N}(\mathbf{x}), \ldots, \nabla_\mathbf{x}R_{N-1,N}(\mathbf{x})] \tag{2.58}$$

$$\nabla_\mathbf{x}R_{n,N}(\mathbf{x}) = \frac{\mathbf{x} - \mathbf{x}_n}{\|\mathbf{x} - \mathbf{x}_n\|^2} - \frac{\mathbf{x} - \mathbf{x}_N}{\|\mathbf{x} - \mathbf{x}_N\|^2} \tag{2.59}$$

In this case, with the Nth measurement as a common reference, the covariance matrix follows (2.13).

2.4.5.3 CRLB for FDOA Measurements

For FDOA, where $\mathbf{z}(\mathbf{x}) = \dot{\mathbf{r}}(\mathbf{x})$ as outlined in (2.16), the Jacobian is given in (12.15) of [1]:

$$\mathbf{J_z}(\mathbf{x}) = \mathbf{J}_{\dot{\mathbf{r}}}(\mathbf{x}) = \left[\nabla_\mathbf{x}\dot{R}_{1,N}(\mathbf{x}), \ldots, \nabla_\mathbf{x}\dot{R}_{N-1,N}(\mathbf{x})\right] \tag{2.60}$$

$$\nabla_\mathbf{x}\dot{R}_{n,N}(\mathbf{x}) = \mathbf{P}_n^\perp(\mathbf{x})\frac{\mathbf{v}_n}{\|\mathbf{x}_n - \mathbf{x}\|^2} - \mathbf{P}_N^\perp(\mathbf{x})\frac{\mathbf{v}_N}{\|\mathbf{x}_N - \mathbf{x}\|^2} \tag{2.61}$$

$$\mathbf{P}_n^\perp(\mathbf{x}) = \mathbf{I} - \frac{(\mathbf{x} - \mathbf{x}_n)(\mathbf{x} - \mathbf{x}_n)^T}{\|\mathbf{x} - \mathbf{x}_n\|^2} \tag{2.62}$$

The covariance matrix follows (2.18). The above was derived for the special case where $\mathbf{v} = \mathbf{0}$. If the target is allowed to have motion, then we must take the gradient with respect to $\mathbf{v}$ as well. In this case, we introduce a slightly expanded notation, where $\mathbf{J}_{\dot{\mathbf{r}},\mathbf{x}}(\mathbf{x},\mathbf{v})$ is the Jacobian with respect to position, and $\mathbf{J}_{\dot{\mathbf{r}},\mathbf{v}}(\mathbf{x})$ is the Jacobian with respect to velocity:

$$\mathbf{J}_{\mathbf{z}}(\mathbf{x},\mathbf{v}) = \begin{bmatrix} \mathbf{J}_{\dot{\mathbf{r}},\mathbf{x}}(\mathbf{x},\mathbf{v}) \\ \mathbf{J}_{\dot{\mathbf{r}},\mathbf{v}}(\mathbf{x}) \end{bmatrix} \tag{2.63}$$

$$= \begin{bmatrix} \nabla_{\mathbf{x}}\dot{R}_{1,N}(\mathbf{x},\mathbf{v}),\ldots,\nabla_{\mathbf{x}}\dot{R}_{N-1,N}(\mathbf{x},\mathbf{v}) \\ \nabla_{\mathbf{v}}\dot{R}_{1,N}(\mathbf{x}),\ldots,\nabla_{\mathbf{v}}\dot{R}_{N-1,N}(\mathbf{x}) \end{bmatrix} \tag{2.64}$$

$$\nabla_{\mathbf{x}}\dot{R}_{n,N}(\mathbf{x},\mathbf{v}) = \mathbf{P}_n^{\perp}(\mathbf{x})\frac{\mathbf{v}_n - \mathbf{v}}{\|\mathbf{x}_n - \mathbf{x}\|^2} - \mathbf{P}_N^{\perp}(\mathbf{x})\frac{\mathbf{v}_N - \mathbf{v}}{\|\mathbf{x}_N - \mathbf{x}\|^2} \tag{2.65}$$

$$\nabla_{\mathbf{v}}\dot{R}_{n,N}(\mathbf{x}) = \frac{\mathbf{x} - \mathbf{x}_n}{\|\mathbf{x} - \mathbf{x}_n\|} - \frac{\mathbf{x} - \mathbf{x}_N}{\|\mathbf{x} - \mathbf{x}_N\|} \tag{2.66}$$

The function `fdoa.jacobian` will return (2.60) when no target velocity is supplied, and (2.63) when it is.

2.4.5.4 CRLB for Hybrid Measurements

For hybrid measurements, the measurement function $\mathbf{z}(\mathbf{x})$ is given in (2.20), which is a stacked version of AOA, TDOA, and FDOA functions. The Jacobian is similarly stacked:

$$\mathbf{J}_{\mathbf{z}}(\mathbf{x}) = \left[\mathbf{J}_{\psi}(\mathbf{x}), \mathbf{J}_{\mathbf{r}}(\mathbf{x}), \mathbf{J}_{\dot{\mathbf{r}}}(\mathbf{x})\right], \tag{2.67}$$

and the covariance matrix follows (2.21).

2.4.6 Calculating CRLB for Geolocation Measurements

In this section, we provide an example to help the reader understand how to compute the CRLB as a prediction of estimation performance and to display the error ellipse that arises from that bound.

Example 2.3 CRLB Calculation for Hybrid Measurements

Consider the scenario in Example 2.2. For the hybrid measurement, with all three types of sensor data, compute the CRLB and display it, compute and display the resultant RMSE, and then overlay the 90% error ellipse for that bound onto plots of the sensor geometry and estimator solutions.

Solution: The first step, computing the CRLB, is quite straightforward, and requires only a call to `computeCRLB` for the appropriate package (`hybrid`, in this case). The result, is a 2×2 covariance matrix.

```
% Compute the CRLB
crlb = hybrid.computeCRLB(x_aoa, x_tdoa, x_fdoa, v_fdoa, x_src, cov_z);
display(crlb);
```

Calculation of the RMSE involves computing the trace of the CRLB, taking the square root, and then (since we're converting the output to km), dividing by 1,000. The CEP50, by comparison, is computed with the function `computeCEP50`, and the error ellipse coordinates are generated with the function `drawErrorEllipse`, both in the `utils` package.

```
% Compute and display the RMSE
fprintf('RMSE: %.2f km\n',sqrt(trace(crlb))/1e3);

% Compute and display the CEP50
cep50 = utils.computeCEP50(crlb);
fprintf('CEP50: %.2f km',cep50/1e3);

% Generate the 90% error ellipse from the CRLB
crlb_ellipse = utils.drawErrorEllipse(x_source, crlb, 101, 90);
```

The output of these computations is shown below, and the error ellipse is drawn in Figure 2.10. This shows that the average error is 210 meters, while the median is 150 meters. The 90% error ellipse is, by comparison, much larger, with a semi-major axis closer to 400 m.

```
crlb =
   1.0e+04 *
    0.5124    0.5082
    0.5082    4.0515

RMSE: 0.21 km
CEP50: 0.15 km
```

□

Figure 2.11 contains a walkthrough of Example 2.3 with additional commentary.

2.5 DISCUSSION

In this chapter, we reviewed three basic aspects of emitter geolocation: the likelihood function that relates measurements to the target's location, classical estimation algorithms that can estimate target location from noisy measurements, and statistical performance prediction, via the Cramér-Rao lower bound. In the following chapters, we will discuss some of the complexities that arise from applying these principles to real-world problems.

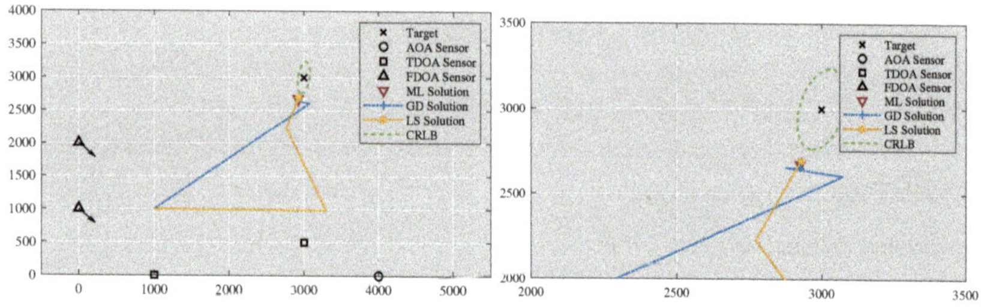

(a) (b)

Figure 2.10 Plot of geolocation solutions overlaid with 90% error ellipse drawn from the Cramér-Rao lower bound for Example 2.3. (a) Full scenario, and (b) close look at target area.

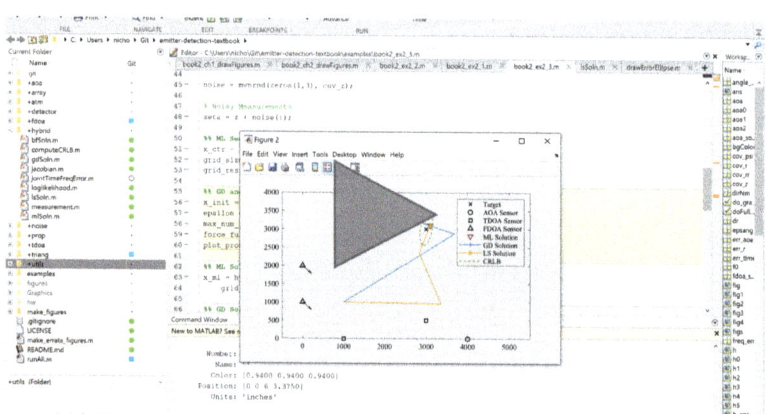

Figure 2.11 Video walkthrough of Example 2.3.

2.6 PROBLEM SET

2.1 Generate the noise-free measurement for a three-sensor DF system with sensors at $(-5, 2)$, $(3, 6)$, and $(0, -2)$, from a target at $(10, 1)$. Generate 1,000 random noise perturbations for that measurement, using a standard deviation of 5° for each sensor. Use the `cov` command to verify that the covariance of the noisy measurements matches what you expect.

2.2 Consider the problem in Example 2.1, but make every sensor a TDOA receiver. Plot the geometry and isochrones using Sensor 4 (0 km, 1 km) as the common reference. Type `help tdoa.measurement` for help defining the reference sensor.

2.3 For the previous problem, generate and display the log likelihood if all of the sensors have a timing accuracy of 1 μs, and again for timing accuracy of 100 ns. Comment on the difference.

2.4 For the previous problem, compute the CEP50 for both timing error cases. Comment on the difference.

2.5 Set up a DF problem with three sensors at (0 km, 0 km), (1 km, 0 km), and (2 km, 0 km), each with an angle error of 5 degrees. Consider a target at (3 km, 5 km). Generate noisy measurements for 1,000 random trials, compute the solution using the three common estimators (maximum likelihood, gradient descent, and iterative least squares), and compute the RMSE for each of the estimators. For the iterative solvers, use an initial position estimate of (1 km, 1 km), and for the the ML solver, search a grid that spans from 0 km to 10 km in both dimensions.

2.6 For the previous problem, compute the CRLB, use it to compute a lower bound on RMSE, and compare to the RMSE for the three estimators in the previous problem.

2.7 Set up a hybrid problem with a pair of TDOA sensors at (1 km, 0 km) and (5 km, 0 km), an AOA sensor at (-1 km, 3 km), and an FDOA sensor pair at (-1 km, 0 km) and (-3 km, 1 km), both moving in the +y direction at 100 m/s. Generate measurements for a target at (2 km, 8 km), emitting a signal at $f_0 = 5$ GHz, with measurement timing accuracy of 100 ns, angle accuracy of 5 degrees, and frequency accuracy of 10 Hz. Use a least square solver to solve for the target position, with an initial position estimate of (3 km, 2 km).

References

[1] N. A. O'Donoughue, *Emitter Detection and Geolocation for Electronic Warfare*. Norwood, MA: Artech House, 2019.

[2] L. L. Scharf, *Statistical Signal Processing*. Reading, MA: Addison-Wesley, 1991.

[3] S. Boyd and L. Vandenberghe, *Convex Optimization*. Cambridge, UK: Cambridge University Press, 2004.

[4] SciPy Documentation, *Correlated random samples*, Aug. 2011. [Online]. Available: https://scipy-cookbook.readthedocs.io/items/CorrelatedRandomSamples.html.

[5] Y.-T. Chan and K. Ho, "A simple and efficient estimator for hyperbolic location," *IEEE Transactions on Signal Processing*, vol. 42, no. 8, pp. 1905–1915, 1994.

[6] K. Ho and Y. Chan, "Geolocation of a known altitude object from TDOA and FDOA measurements," *IEEE Transactions on Aerospace and Electronic Systems*, vol. 33, no. 3, pp. 770–783, 1997. DOI: 10.1109/7.599239.

[7] F. Quo and K. Ho, "A quadratic constraint solution method for TDOA and FDOA localization," in *2011 IEEE International Conference on Acoustics, Speech and Signal Processing (ICASSP)*, IEEE, 2011, pp. 2588–2591.

[8] L. Lin, H. So, F. K. Chan, Y. Chan, and K. Ho, "A new constrained weighted least squares algorithm for TDOA-based localization," *Signal Processing*, vol. 93, no. 11, pp. 2872–2878, 2013, ISSN: 0165-1684.

[9] Y. Wang and K. Ho, "TDOA positioning irrespective of source range," *IEEE Transactions on Signal Processing*, vol. 65, no. 6, pp. 1447–1460, 2017.

[10] Y. Wang and K. C. Ho, "Unified near-field and far-field localization for AOA and hybrid AOA-TDOA positionings," *IEEE Transactions on Wireless Communications*, vol. 17, no. 2, pp. 1242–1254, Feb. 2018, ISSN: 1536-1276. DOI: 10.1109/TWC.2017.2777457.

[11] R. L. Elder, "An examination of circular error probable approximation techniques," Air Force Institute of Technology, School of Engineering, Wright-Patterson AFB, OH, Tech. Rep., 1986.

[12] RAND Corporation, "Offset circle probabilities," RAND Corporation, Tech. Rep., 1952.

[13] L. H. Wegner, "On the accuracy analysis of airborne techniques for passively locating electromagnetic emitters," RAND, Santa Monica, CA, Tech. Rep., 1971.

[14] W. Nelson, "Use of circular error probability in target detection," MITRE, Bedford, MA, Tech. Rep., 1988.

[15] National Research Council, *The global positioning system: A shared national asset*. Washington, D.C.: National Academies Press, 1995.

[16] D. R. Childs, D. M. Coffey, and S. P. Travis, "Error statistics for normal random variables," Naval Underwater Systems Center, Newport, RI, Tech. Rep., 1975.

[17] H. L. Van Trees, *Detection, Estimation, and Linear Modulation: Part I of Detection, Estimation, and Modulation Theory*. Hoboken, NJ: Wiley, 2001.

[18] S. M. Kay, *Fundamentals of Statistical Signal Processing, Volume 1: Estimation Theory*. Englewood Cliffs, NJ: PTR Prentice-Hall, 1993.

[19] A. Nehorai and G. Tang, "Bounds on estimation," in *Encyclopedia of Systems and Control*, J. Baillieul and T. Samad, Eds. London: Springer London, 2015, pp. 93–98, ISBN: 978-1-4471-5058-9. DOI: 10.1007/978-1-4471-5058-9_69. [Online]. Available: https://doi.org/10.1007/978-1-4471-5058-9_69.

[20] P. J. Schreier and L. L. Scharf, *Statistical Signal Processing of Complex-Valued Data: The Theory of Improper and Noncircular Signals*. Cambridge, UK: Cambridge University Press, 2010.

[21] E. Barankin, "Locally best unbiased estimates," *The Annals of Mathematical Statistics*, pp. 477–501, 1949.

[22] R. McAulay and E. Hofstetter, "Barankin bounds on parameter estimation," *IEEE Transactions on Information Theory*, vol. 17, no. 6, pp. 669–676, 1971. DOI: 10.1109/TIT.1971.1054719.

[23] A. V. Fend, "On the attainment of Cramer-Rao and Bhattacharyya bounds for the variance of an estimate," *The Annals of Mathematical Statistics*, pp. 381–388, 1959.

[24] J. S. Abel, "A bound on mean-square-estimate error," *IEEE Transactions on Information Theory*, vol. 39, no. 5, pp. 1675–1680, Sep. 1993, ISSN: 0018-9448. DOI: 10.1109/18.259655.

[25] J. M. Hammersley, "On estimating restricted parameters," *Journal of the Royal Statistical Society. Series B (Methodological)*, vol. 12, no. 2, pp. 192–240, 1950.

[26] D. G. Chapman and H. Robbins, "Minimum variance estimation without regularity assumptions," *Annals of Mathematical Statistics*, vol. 22, no. 4, pp. 581–586, 1951.

[27] H. L. Van Trees and K. L. Bell, "A bound on meansquareestimate error," in *Bayesian Bounds for Parameter Estimation and Nonlinear Filtering/Tracking*. John Wiley, 2007, pp. 171–175. DOI: 10.1109/9780470544198.ch11.

[28] J. Ziv and M. Zakai, "Some lower bounds on signal parameter estimation," *IEEE Transactions on Information Theory*, vol. 15, no. 3, pp. 386–391, May 1969, ISSN: 0018-9448. DOI: 10.1109/TIT.1969.1054301.

[29] K. L. Bell, Y. Steinberg, Y. Ephraim, and H. L. Van Trees, "Extended Ziv-Zakai lower bound for vector parameter estimation," *IEEE Transactions on Information Theory*, vol. 43, no. 2, pp. 624–637, Mar. 1997, ISSN: 0018-9448. DOI: 10.1109/18.556118.

[30] N. Decarli and D. Dardari, "Ziv-Zakai bound for time delay estimation of unknown deterministic signals," in *2014 IEEE International Conference on Acoustics, Speech and Signal Processing (ICASSP)*, May 2014, pp. 4673–4677. DOI: 10.1109/ICASSP.2014.6854488.

[31] D. Khan and K. L. Bell, "Explicit Ziv-Zakai bound for analysis of DOA estimation performance of sparse linear arrays," *Signal Processing*, vol. 93, no. 12, pp. 3449–3458, 2013, ISSN: 0165-1684.

[32] D. Dardari, A. Conti, U. Ferner, A. Giorgetti, and M. Z. Win, "Ranging with ultrawide bandwidth signals in multipath environments," *Proceedings of the IEEE*, vol. 97, no. 2, pp. 404–426, Feb. 2009, ISSN: 0018-9219. DOI: 10.1109/JPROC.2008.2008846.

Chapter 3

Sensor Selection

3.1 INTRODUCTION

When using time or frequency *difference* as a geolocation measurement, there is a fundamental requirement to compare the measurement at two different sensors. This is because the time (and carrier frequency) of the original transmission are assumed to be unknown. Many popular formulations for TDOA and FDOA systems use a common reference sensor for all measurements, such as [1]–[10]. This scenario is illustrated in Figure 3.1. This is referred to in the literature as a *nonredundant set*, or in a few cases as the *spherical TDOA set*, and results in N-1 sensor pair measurements for any set of N sensors. Information theory suggests that this should be sufficient, since it contains information from all of the available sensors. In several studies, such as [11], it has been found that the choice of reference sensor for TDOA may also be irrelevant, in that it does not affect the CRLB.

Some studies have found a small but noticeable improvement in the CRLB in cases where all of the N(N-1)/2 possible unique sensor pairs were considered, which one study suggests is on the order of N (via analysis of the *geometric dilution of precision*) [12], [13]. Figure 3.2 illustrates this scenario. The full set has been used in several geolocation papers, a small sample of which is cited here [14]–[17].

Many authors have sought to optimally select a nonredundant set, in order to minimize the amount of processing or data transmission required, without sacrificing geolocation performance [18]–[20]. In one such study, they found slight improvements in performance for the full set vs. a nonoptimized nonredundant set, but were able to match its performance with an optimized nonredundant set, suggesting that intelligent selection of sensor pairs is possible without loss of geolocation performance [18].

In a more fundamental theoretical analysis, Velasco et al. have analyzed the properties of the *TDOA matrix*, the N $\times$ N matrix of all possible TDOA pairs in a given network of N sensors [21]. Through some fundamental linear algebra, it is shown that the TDOA matrix has rank 2, proving that the vast majority of its time difference

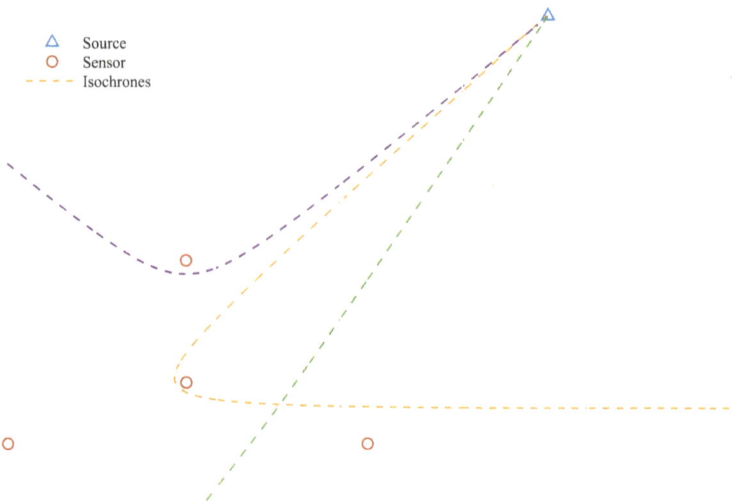

Figure 3.1 Illustration of a common reference sensor in a multisite TDOA system.

measurements are redundant. This low rank matches the conclusion that not all N(N-1)/2 unique time difference measurements are required and the inference above that as few as (N-1) measurements contains all of the required information for geolocation. However, the authors were studying measurement sets in the context of denoising algorithms, and rejection of outliers. In this situation, they found clear benefit to keeping the full measurement set, as it greatly improved the performance of denoising and outlier rejection.

Another driving factor in sensor selection is communications and processing throughput. If individual sensors compute time/frequency of arrival estimates, then coordination of those estimates is a fairly low-overhead operation (less than a thousand bits to transmit an estimate and the necessary pulse descriptor word to allow for comparison with other estimates), and the processing demand of considering the extra difference measurements in the geolocation stage of the algorithm is relatively minor (the most expensive step is the inversion of a larger covariance matrix). However, if the more sensitive *correlation*-based time and frequency difference estimates are needed, then complex (I&Q) data must be transmitted for each intercepted pulse, and to achieve performance against weak signals (that may not be visible except via correlation processing), large chips must be sent for correlation processing at a central node. This easily increases the data throughput requirements to hundreds of megabits per second (Mbps) or even gigabits per second (Gbps). Such high-throughput communications may not be available for all sensors. As such, the selection of which sensor pairs are compared

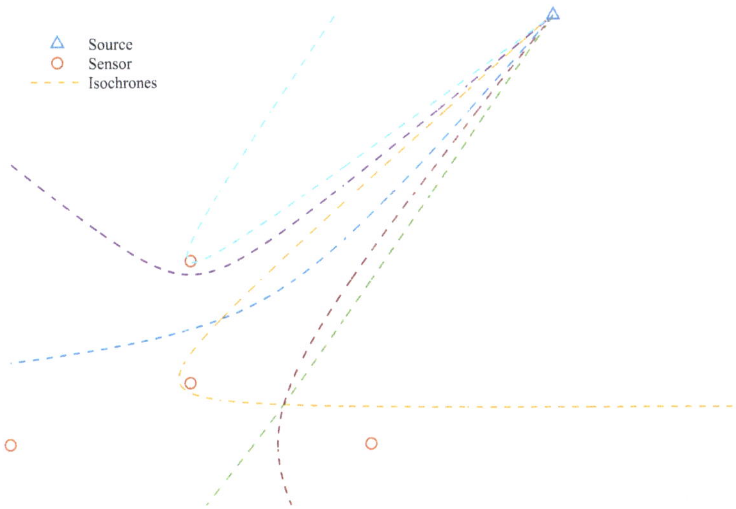

Figure 3.2 Illustration of a complete reference set in a multisite TDOA system.

to generate the TDOA/FDOA estimates for geolocation may be dictated by geography or topology of the available sensors, rather than by some optimization algorithm.

In this chapter, we will investigate both full and nonredundant sensor sets, and present a utility in the provided MATLAB® code that allows for definition of arbitrary sensor sets, so that users are able to experiment with various sensor selection strategies. We will also briefly discuss, but not derive or analyze, a few representative optimization approaches found in the literature.

3.2 MEASUREMENT SETS

In this section, we describe the two broad classes of measurement sets: those that leverage all N(N-1)/2 unique sensor pairs (*full measurement sets*), and those that use only (N-1) sensor pairs (*nonredundant sets*).

3.2.1 Full Measurement Set

For any set of N sensors, the full measurement set is straightforward to derive or visualize, but can be quite computationally intensive, particularly if the underlying estimator is a cross-correlation TDOA estimate or a complex ambiguity function-based TDOA/FDOA estimate. The benefit of this extra computation is optimal performance, in terms of geometric and statistical stability, as the best possible information is accounted for in every scenario, although it is conceivable that the same performance could be achieved with fewer sensor pairs.

The structure of $\widetilde{\mathbf{C}}$ can be complex, however, even under the ideal assumption that all sensor measurements are independent. Since the information from each sensor will appear in N-1 of the sensor pairs, each of the measurements will have some correlation with 2N-4 other sensor pairs (N-2 for the first, N-2 for the second). Thus, each row or column will have 2N-3 non-zero entries (including the autocorrelation term on the principal diagonal).

As an example, consider a four sensor setup with noise variances $\sigma_0^2, \dots, \sigma_3^2$, and independence between the measurements. The indices for the full sensor measurement set are:

$$\mathbf{a} = [0, 0, 0, 1, 1, 2], \tag{3.1}$$

$$\mathbf{b} = [1, 2, 3, 2, 3, 3], \tag{3.2}$$

where a_i and b_i together form the ith sensor pair. With these sensor pairs, the covariance matrix takes the form:

$$\begin{bmatrix} \sigma_0^2+\sigma_1^2 & \sigma_0^2 & \sigma_0^2 & \sigma_1^2 & \sigma_1^2 & 0 \\ \sigma_0^2 & \sigma_0^2+\sigma_2^2 & \sigma_0^2 & \sigma_2^2 & 0 & \sigma_2^2 \\ \sigma_0^2 & \sigma_0^2 & \sigma_0^2+\sigma_3^2 & 0 & \sigma_3^2 & \sigma_3^2 \\ \sigma_1^2 & \sigma_2^2 & 0 & \sigma_1^2+\sigma_2^2 & \sigma_1^2 & \sigma_2^2 \\ \sigma_1^2 & 0 & \sigma_3^2 & \sigma_1^2 & \sigma_1^2+\sigma_3^2 & \sigma_3^2 \\ 0 & \sigma_2^2 & \sigma_3^2 & \sigma_2^2 & \sigma_3^2 & \sigma_2^2+\sigma_3^2 \end{bmatrix} \tag{3.3}$$

For simplicity (and illustration), let us consider the case where all sensors have equal measurement error. The structure of the covariance matrix is:

$$\begin{bmatrix} 2 & 1 & 1 & 1 & 1 & 0 \\ 1 & 2 & 1 & 1 & 0 & 1 \\ 1 & 1 & 2 & 0 & 1 & 1 \\ 1 & 1 & 0 & 2 & 1 & 1 \\ 1 & 0 & 1 & 1 & 2 & 1 \\ 0 & 1 & 1 & 1 & 1 & 2 \end{bmatrix} \sigma^2 \tag{3.4}$$

From the discussion above, we know that each row or column with have 2N-3 (for N=4, the result is 5) non-zero entries in each row or column, which is observed here. As the number of sensors grows, there will be more non-zero entries in each row or column. The off-diagonal entries in this matrix represent scenarios where a given sensor is used in multiple sensor pairs.

3.2.2 Nonredundant Set

A nonredundant set is any set that does not contain measurements that could have been reconstructed from other measurements in the set. This is similar to the concept of a matrix whose columns are linearly independent [22]. To illustrate this, consider three sensor pair measurements:

$$\zeta_{1,0} = \zeta_1 - \zeta_0 \tag{3.5}$$

$$\zeta_{2,1} = \zeta_2 - \zeta_1 \tag{3.6}$$

$$\zeta_{2,0} = \zeta_2 - \zeta_0 = \zeta_{2,1} + \zeta_{1,0} \tag{3.7}$$

Since the third measurement pair $\zeta_{2,0}$ is a function of the first two, this set is redundant. The advantage of a nonredundant set is that it represents all available information with a minimal set of data, minimizing computational overhead.

The simplest approach to derive a nonredundant set is the use of a common reference sensor. In many cases, the selection is irrelevant, particularly if the target is at long range and all sensors have similar SNR.

3.2.2.1 Optimal Selection

Several researchers have derived methods for optimal selection of sensors, in order to minimize computational overhead without sacrificing performance, with respect to the full measurement set. It has been shown that these optimal sets can match the performance of the full measurement set [19], [20].

The challenge, in EW scenarios, is that the target of interest is often at very long range and, thus, low SNR. The optimal approach for detecting these very challenging signals is via the cross-correlation (for TDOA) or complex ambiguity function (for TDOA/FDOA). These steps are computationally complex and often present the most costly step for geolocation. So, it would be ideal to perform optimal sensor selection before the cross-correlation step. Unfortunately, optimal sensor selection is often dependent on the specifics of a target, such as the relative SNR at each sensor, or even just confidence that each sensor has detected the target signal.

This is not to say that there is no value in optimal sensor selection, but that care must be taken to consider the relative computation (and communications) benefits when deriving a sensor selection strategy. For instance, assigning pair-wise sensors to do initial cross-correlation could present a suboptimal initial set of sensors, and then could be used to select which additional pairs are ideal for processing.

Detailed analysis of specific sensor selection strategies is outside the scope of this book, but is an interesting area of research for further development.

3.3 MEASUREMENT STATISTICS OF SENSOR PAIRS

In order to understand the performance of geolocation with arbitrary sensor pairs, let us discuss their measurement statistics. Let $z_{i,j}$ be the noise-free parameter found by comparing sensor i and sensor j, and $\zeta_{i,j}$ be the noisy version, while z_i and ζ_i are the underlying single sensor measurements.[1.] We define these measurements as Gaussian random variables with mean value equivalent to the underlying parameters and some noise power σ_i^2. It is straightforward to express the joint measurement similarly:

$$\zeta_i \sim \mathcal{N}\left(z_i, \sigma_i^2\right) \tag{3.8}$$

$$\zeta_j \sim \mathcal{N}\left(z_j, \sigma_j^2\right) \tag{3.9}$$

$$z_{i,j} \triangleq z_j - z_i \tag{3.10}$$

$$\zeta_{i,j} \triangleq \zeta_j - \zeta_i \mathcal{N}\left(z_{i,j}, \sigma_{i,j}^2\right) \tag{3.11}$$

Now, if ζ_i and ζ_j are independent measurements, then the difference has variance defined simply by the sum of their component variances ($\sigma_{i,j}^2 = \sigma_i^2 + \sigma_j^2$) [22]. If, however, they are correlated, then the variance of their difference is given:

$$\sigma_{i,j}^2 \triangleq E\left\{\left(\zeta_{i,j} - z_{i,j}\right)^2\right\} \tag{3.12}$$

$$= \sigma_i^2 + \sigma_j^2 - 2E\left\{(\zeta_i - z_i)(\zeta_j - z_j)\right\} \tag{3.13}$$

$$= \sigma_i^2 + \sigma_j^2 - 2\rho_{i,j}\sigma_i\sigma_j \tag{3.14}$$

where $\rho_{i,j}$ is the *correlation coefficient* of the two measurements.

3.3.1 Sensor Pair Cross-Covariance

Let us consider two arbitrary sensor pairs $z_{i,j}$ and $z_{k,l}$. Both are Gaussian random variables, as discussed above, but they are also *jointly Gaussian*.[2.] In this case, their joint distribution is completely characterized by their means and variances, derived in (3.14), as well as by their cross-covariance. For this, we compute the expectation

$$E\left\{\left(\zeta_{i,j} - z_{i,j}\right)\left(\zeta_{k,l} - z_{k,l}\right)\right\} = \rho_{i,k}\sigma_i\sigma_k + \rho_{j,l}\sigma_j\sigma_l - \rho_{j,k}\sigma_j\sigma_k - \rho_{i,l}\sigma_i\sigma_l \tag{3.15}$$

Let us assume that the covariance matrix $\mathbf{C}$ has been provided for the individual sensor measurements, and define the i, jth element

$$\left[\mathbf{C}\right]_{i,j} \triangleq \rho_{i,j}\sigma_i\sigma_j \tag{3.16}$$

1. In some cases, such as correlation-based TDOA or joint time/frequency difference estimation via the complex ambiguity function (CAF), there are no single sensor measurements, but $\mathbf{z}_i$ would be the ideal (noise-free) input signal from the ith sensor, and $z_{i,j}$ would be the cross-correlation derived output from comparing the ith and jth sensors.
2. Both $z_{i,j}$ and $z_{k,l}$ are affine transformations of the measurements ζ_i, ζ_j, ζ_k, ζ_l, which we assume are themselves *jointly Gaussian*, meaning that the sensor pairs (indeed, all sensor pairs) are themselves jointly Gaussian [22].

We can then express the cross-covariance in matrix terms:

$$E\{(\zeta_{i,j} - z_{i,j})(\zeta_{k,l} - z_{k,l})\} = [\mathbf{C}]_{i,k} + [\mathbf{C}]_{j,l} - [\mathbf{C}]_{j,k} - [\mathbf{C}]_{i,l} \tag{3.17}$$

In this manner, we can build the covariance matrix for a set of arbitrary sensor pairs. Let $\mathbf{a} = [a_0, \dots, a_{L-1}]$ be a set of reference sensor indices, and $\mathbf{b} = [b_0, \dots, b_{L-1}]$ be a set of test sensor indices. The lth measurement is given by ζ_{a_l,b_l}. We use the notation $\widetilde{\mathbf{C}}$ to denote the covariance matrix of the sensor pairs, while $\mathbf{C}$ is the covariance matrix of the component measurements that are available. To build $\widetilde{\mathbf{C}}$, we express the (i,j)th element of the covariance matrix:

$$\left[\widetilde{\mathbf{C}}\right]_{i,j} = E\left\{(\zeta_{a_i,b_i} - z_{a_i,b_i})\left(\zeta_{a_j,b_j} - z_{a_j,b_j}\right)\right\} \tag{3.18}$$

$$= [\mathbf{C}]_{a_i,a_j} + [\mathbf{C}]_{b_i,b_j} - [\mathbf{C}]_{a_i,b_j} - [\mathbf{C}]_{a_j,b_i} \tag{3.19}$$

The diagonal elements (where $i = j$) simplify to:

$$\left[\widetilde{\mathbf{C}}\right]_{i,j} = [\mathbf{C}]_{a_i,a_i} + [\mathbf{C}]_{b_i,b_i} - 2\,[\mathbf{C}]_{a_i,b_i} \tag{3.20}$$

3.3.2 Usage

The functionality to convert from $\mathbf{C}$ to $\widetilde{\mathbf{C}}$ is provided in the function `parseReferenceSensor` and `resampleCovarianceMatrix` of the `utils` package. The MATLAB® code for all TDOA, FDOA, and hybrid solvers, performance estimators, and measurement generators has been modified to call these functions.

Previously, the MATLAB® code was written to accept a reference index for the common reference sensor. That parameter now has multiple possible ways to be used. It can be used as originally defined (a scalar index to the reference sensor), or it can be a 2 × N vector of sensor pair indices, for any arbitrary set of sensor pairs, or it can be a string with the word `'full'` (case insensitive) to utilize the full N(N-1)/2 set of unique sensor pairs. This is handled by `parseReferenceSensor`

```
> n_sensors = 10;
> ref_idx = 7;
> [test_idx_vec, ref_idx_vec] = utils.parseReferenceSensor(ref_idx,
    n_sensors);
```

In a slightly more complex case, we have already defined a set of sensor measurement pairs as a 2 × N matrix:

```
> n_sensors = 10;
> ref_idx = [1 2; 3 4; 5 8; 6 9; 7 10]';
```

```
> [test_idx_vec, ref_idx_vec] = utils.parseReferenceSensor(ref_idx,
    n_sensors);
test_idx_vec = [1  3  5  6  7]
ref_idx_vec = [2 4 8 9 10]
```

Finally, to obtain the full set of measurements, set the reference index input to `'full'`.

```
> n_sensors = 4;
> ref_idx = 'full';
> [test_idx_vec, ref_idx_vec] = utils.parseReferenceSensor(ref_idx,
    n_sensors);
test_idx_vec = [1 1 1 2 2 3]
ref_idx_vec = [2 3 4 3 4 4]
```

In each case, the function returns a matching pair of sensor indices which can then be used to generate measurements. Users don't have to directly call this function if they don't want to, however. It is implicitly called by all of the relevant measurement and performance predictions, as well as by the `resampleCovarianceMatrix` utility that executes the covariance matrix resampling discussed in the previous section.

```
> C_tilde = utils.resampleCovMtx(C, ref_idx);
```

The result, `C_tilde` is an `n_pair` by `n_pair` covariance matrix for all of the commanded sensor pairs. This works for TDOA/FDOA pairs, so long as they are stacked properly (TDOA in the upper-left quadrant and FDOA in the lower-right), with sensor indices defined properly. However, it is slightly more complex if AOA is used in addition to TDOA and FDOA, since it does not rely on sensor pairs the way that the other two do. To handle this, define the reference index for all AOA sensors with `NaN`. For example, consider the following covariance matrix built from three sensors, each capable of AOA, TDOA, and FDOA. In this simple scenario, all three AOA sensors have measurement error `ang_err`, all three TDOA sensors have range of arrival error `rng_err`, and all three FDOA sensors have range-rate of arrival error `rng_rate_err`. For simplicity, we'll assume that all measurements are independent. The full measurement covariance matrix is the block diagonal combination of the three components.

```
> C_aoa = ang_err^2*eye(3);
> C_roa = rng_err^2*eye(3);
> C_rroa = rng_rate_err^2*eye(3);
> C_full = blkdiag(C_aoa, C_roa, C_rroa);
```

Next, we define the references for each type of measurement. Let's assume that all three AOA sensors are used, and that both TDOA and FDOA will use the first as a common reference (note that the first TDOA sensor is index 4, while the first FDOA sensor is index 7). We can combine them into a $2 \times N$ matrix for passing in, or we have the freedom to directly pass the two vectors to `resampleCovMtx`, as shown in the following snippet.

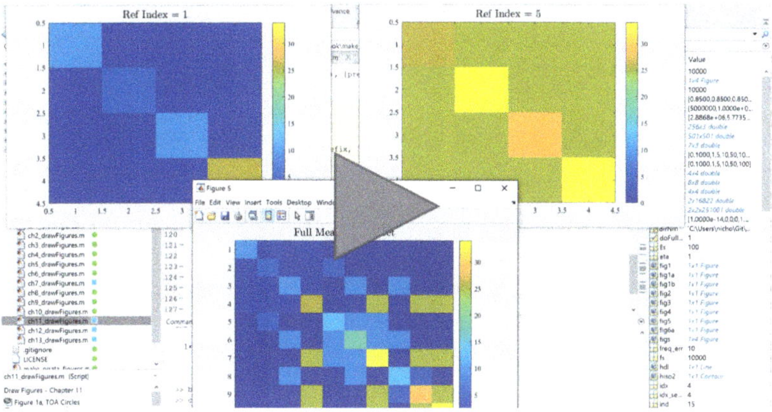

Figure 3.3 Video discussion of sensor pair covariance.

```
test_idx_vec = [1, 2, 3, 5, 6, 8, 9];
ref_idx_vec = [NaN, NaN, NaN, 4, 4, 7, 7];
C_tilde = utils.resampleCovMtx(C_full, test_idx_vec, ref_idx_vec);
```

If we plan to call any of the provided measurement, solver, or performance prediction utilities, we'll need to combine them into a single matrix for passing.

```
ref_idx = [test_idx_vec; ref_idx_vec]; % Combine into 2 x N matrix
```

Figure 3.3 contains a link to a video describing sensor pair covariance, how the matrices are generated, and how to resample noise to match the desired covariance for a specific set of sensor pairs, including a walkthrough of Example 3.1.

Example 3.1 Covariance Matrix Resampling

Consider a five-sensor TDOA geometry, with independent sensor measurements and range-of-arrival error that is zero-mean Gaussian distributed with standard deviations of 1, 3, 2, 3, and 5 m, respectively. Generate the covariance matrix for TDOA measurements (in terms of m^2) using the first sensor as a reference, and again using the final sensor as a reference. Compare. Then, generate the covariance matrix for the full measurement set and comment on the differences.

Solution: To begin, we'll assemble the base covariance matrix, which is a diagonal matrix given the independence of the measurements.

```
C = diag([1, 3, 2, 3, 5].^2); % Square to convert from std dev to variance
```

Next, we generate the covariance matrices for the two common reference cases.

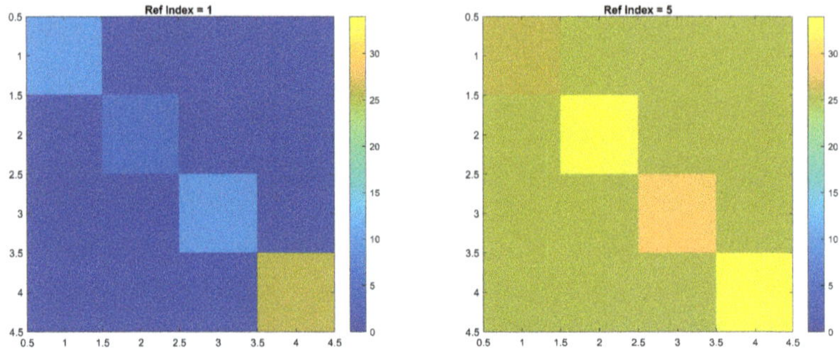

(a) (b)

Figure 3.4 RDOA covariance matrices for (a) first and (b) last sensors as the common reference sensor, from Example 3.1.

```
C_first = utils.resampleCovMtx(C, 1);
C_last = utils.resampleCovMtx(C, 5);
```

We plot the results in Figure 3.4. We see in both cases that the covariance matrix has a raised diagonal and uniform (non-zero) off-diagonal entries, from the common reference sensor. This matches the structure described in Chapters 11–13 of [23]. Also, we see significantly higher off-diagonal entries in Figure 3.4(b), due to the higher variance for the common reference sensor used in this case. This highlights the importance of reference sensor selection when SNR is not uniform.

Next, we generate the full measurement set.

```
C_full = utils.resampleCovMtx(C, 'full');
```

The result is printed in Figure 3.5. In this case, the covariance matrix is much larger, and has a structure to its off-diagonal entries, since there isn't a single common reference sensor that is used everywhere, but each sensor is used N-1 times. □

3.3.3 Generating Measurements

The provided tools have been upgraded to ensure consistent use of the reference indices. For instance, to generate measurements for a hybrid scenario, one can supply two sets of indices to define how the TDOA and FDOA sensor pairs are assigned. Originally, this was required to be a single integer (indicating the position of the common reference), but now they support any of the allowable inputs to `parseReferenceSensor`. If not provided, the default behavior remains to use the Nth sensor as a common reference.

For example, to define a complete measurement set, use the commands:

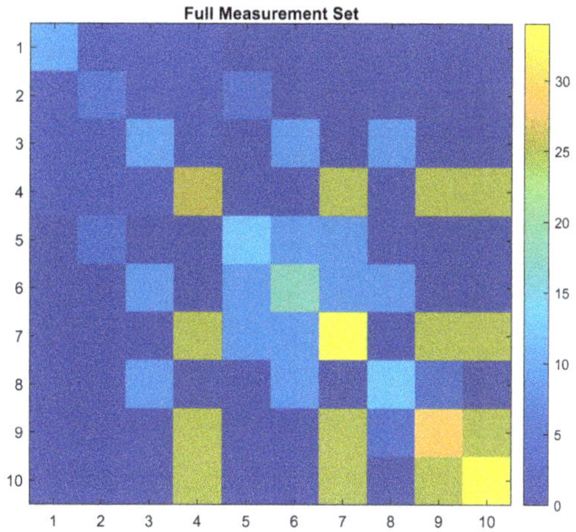

Figure 3.5 RDOA covariance matrix for the full measurement set, from Example 3.1.

```
> tdoa_ref = 'full';
> fdoa_ref = 'full';
> z = hybrid.measurement(x_aoa, x_tdoa, x_fdoa, v_fdoa, x_source, C_full,
    tdoa_ref, fdoa_ref);
```

To add noise to these measurements, we need to consider the covariance matrix $\tilde{\mathbf{C}}$. Users with the *Statistics & Machine Learning Toolbox* can call the built-in MATLAB® function `mvn` to generate a multivariate normal distribution with any covariance matrix. To do so manually, we can leverage the Cholesky decomposition [22]. We briefly discussed this in Example 2.2, but will now more clearly explore why this works.

Any Hermitian positive semidefinite matrix, such as a covariance matrix, can be decomposed into a lower-left triangular matrix $\mathbf{L}$, or upper-right triangular matrix $\mathbf{R}$, such that[3].

$$\tilde{\mathbf{C}} = \mathbf{L}\mathbf{L}^H = \mathbf{R}^H\mathbf{R} \tag{3.21}$$

The advantage to this decomposition is that we can take a set of independent zero-mean Gaussian samples $\mathbf{y}$ (with covariance matrix $\mathbf{I}$) and premultiply them by $\mathbf{L}$ to generate correlated measurements with covariance matrix $\mathbf{L}\mathbf{L}^H$. To confirm, consider

3. Since $\tilde{\mathbf{C}}$ is real-valued, the $\cdot^H$ operators can be replaced with $\cdot^T$.

the covariance of $\mathbf{Ly}$.

$$E\left\{(\mathbf{Ly})(\mathbf{Ly})^T\right\} = E\left\{\mathbf{Lyy}^T\mathbf{L}^T\right\} \tag{3.22}$$
$$= \mathbf{L}E\left\{\mathbf{yy}^T\right\}\mathbf{L}^T \tag{3.23}$$
$$= \mathbf{LIL}^T = \tilde{\mathbf{C}} \tag{3.24}$$

Thus, we see that by premultiplying an ideal (uniform and independent) set of standard Gaussian samples $\mathbf{y}$ by the lower-left Cholesky decomposition of $\tilde{\mathbf{C}}$, we can generate correlated samples that have covariance $\tilde{\mathbf{C}}$. Written in MATLAB® code, this is achieved with a few commands.

```
L = chol(C_tilde, 'lower');
num_z = size(C_tilde, 1);
num_samples = size(z, 2);
noise = L * randn(num_z, num_samples);
zeta = noise + z;
```

Alternatively, the noise measurements can be generated at a sensor level, and then resampled, just like the covariance matrix, using the function `resampleNoise`.

```
noise_sensor = randn(num_sensors, 1);
noise = utils.resampleNoise(noise_sensor, test_idx, ref_idx);
```

The result is equivalent using these two methods, but the advantage of resampling the noise is that a single set of sensor-level noise measurements can be generated, and then resampled for various sensor selection schemes, so that different sensor selections are using the same underlying noise, for direct comparisons of performance and results.

3.3.4 Jacobian Matrix of Arbitrary Sensor Pairs

For a given set of sensor pairs $\mathbf{a}$ and $\mathbf{b}$, defined above, the Jacobian matrix is generated by simply selecting different range and range-rate gradients from the full set of possible pairs:

$$\mathbf{J_z}(\mathbf{x}) = \left[\nabla_{\mathbf{x}} R_{a_0,b_0}, \dots, \nabla_{\mathbf{x}} R_{a_{N-1},b_{N-1}}\right] \qquad \text{for TDOA} \tag{3.25}$$
$$= \left[\nabla_{\mathbf{x}} \dot{R}_{a_0,b_0}, \dots, \nabla_{\mathbf{x}} \dot{R}_{a_{N-1},b_{N-1}}\right] \qquad \text{for FDOA} \tag{3.26}$$

Similar to the covariance matrix resampler, the Jacobian generator accepts an argument for the sensor selection indices and returns the Jacobian. This can be called in the `tdoa`, `fdoa`, and `hybrid` packages.

```
> J = tdoa.jacobian(x_source, x_sensor, ref_idx)
```

The code for computing the CRLB accepts these sensor indices and passes on to the covariance matrix resampler and Jacobian matrix generator.

```
> CRLB = tdoa.crlb(x_sensor, x_source, C, ref_idx)
```

In some limited testing, we have found that the matrices generated by `resampleCovMtx` are not always positive semidefinite (a requirement for matrix inversion, and a theoretically guaranteed feature of covariance matrices). This is a common practical occurrence in numerical simulations and sample covariance matrix calculations, and typically arises from rounding errors, which can have the effect of shifting an eigenvalue that was zero (or very close to it) to appear to be less than zero. This causes errors during the calculation of the CRLB or of the log likelihood, as matrix inversion requires that all eigenvalues be nonnegative (known as a *positive semidefinite* matrix). Thus, we've implemented a utility called `ensureInvertible`, which applies a diagonal loading term to ensure that the matrix is positive semidefinite (see [24] for an early discussion of diagonal loading to increase stability of sample covariance matrices).

3.4 PERFORMANCE ANALYSIS

This section is not an exhaustive comparison of full and nonredundant sets. We accept the conclusions of those researchers who have found it possible to select an optimized nonredundant set. Rather, we seek to establish some test scenarios, and demonstrate the tools and methods available to compare different sensor pair sets in a given scenario. We will first compare the performance via the CRLB in two different scenarios, then examine the behavior of implemented solvers in a third test.

Example 3.2 TDOA Sensor Selection Example

First, we compare a very straightforward setup with four TDOA sensors in a Y-configuration, with a 10-km baseline, and consider targets as far as 100 km away. For simplicity, we operate in a 2-D space. We will compare four different nonredundant sets (one for each sensor designated as the common reference) to the full set. Our metric for performance will be the lower bound on the CEP50, computed from the CRLB. For this test, we will assume a uniform timing error of 100 ns across all of the TDOA sensors.

Solution: We begin by defining, and plotting, the sensor positions. For brevity, we omit the code used to define these positions. The geometry is plotted in Figure 3.6.

Next, we define the covariance matrix. The function `tdoa.computeCRLB` is built to accept a covariance matrix in units of time (s^2), so we define the covariance matrix directly with the specified timing error.

```
time_err = 100e-9; % 100 ns
cov_full = time_err^2 * eye(n_sensors);
```

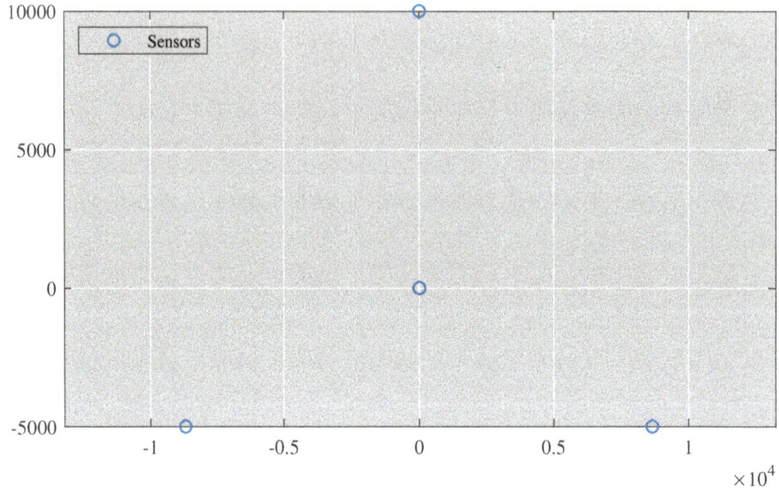

Figure 3.6 Geometry for Example 3.2.

Next, we define the different sensor pairs. For ease of coding later, we'll store them in a cell array; the first four entries step through each sensor as the common reference. Finally, we use the `'full'` shorthand keyword for the full measurement set.

```
ref_set = {1,2,3,4,'full'};
```

Finally, we execute our for loop across the different sensor pair options. With each option, we'll call the `computeCRLB` function to compute the CRLB at each of the test points stored in `x_source`, and then compute the CEP50 for each test point.

```
for idx_set = 1:numel(ref_set)
    this_ref = ref_set{idx_set};

    % Repeat x_sensors for both the TDOA and FDOA sensor sets, because
    % all sensors do both.  Repeat the ref index, because the sensor
    % pairs for TDOA and FDOA are the same.
    this_crlb = hybrid.computeCRLB([],x_sensor, x_sensor, v_sensor, ...
                                   x_source, cov_full, this_ref, ...
                                   this_ref);

    this_cep = utils.computeCEP50(this_crlb);

    % Plot this result
    ...
end
```

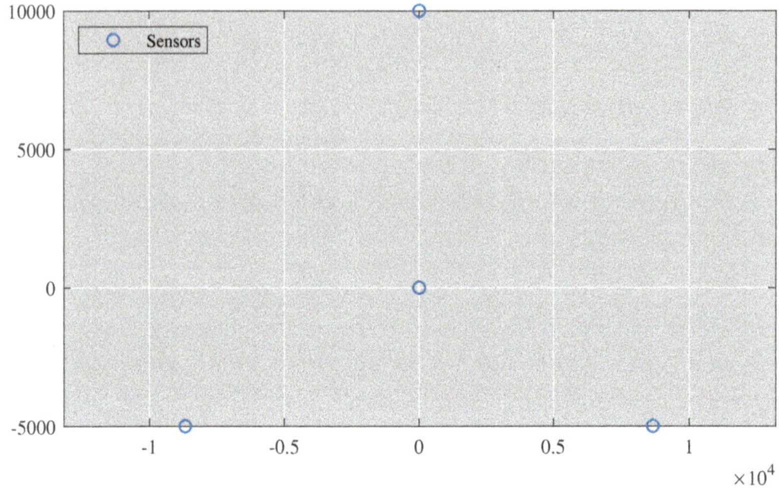

The results are plotted in Figure 3.7 for all five sensor pair configurations. The obvious result is that there is no discernible difference between any sensor pair option, including the full measurement set. This example has demonstrated how to easily test different sensor selections via the CRLB. □

Figure 3.8 contains a brief walkthrough of Example 3.2 and adds consideration of a third dimension.

Example 3.3 Four-Ship of TDOA/FDOA Sensors

In this example, let's consider a four-ship of airborne sensors performing joint TDOA/F-DOA. The sensors are configured as two-ships, spaced at 5 km, with 25 km between the two-ships, all along the x-axis. The left two-ship is moving at 100 m/s, at an angle of 80 degrees (CCW from the +x axis), and the right two-ship is moving at 100 m/s, at an angle of 100 degrees (CCW from the +x axis). Targets are expected up to 50 km away from the center of the formation, in any direction. Assume the sensors and targets are coaltitude, so that we can conduct our analysis in 2-D.

We will compare the geolocation performance, again via the CEP50 computed from the CRLB, this time for three different sensor pair configurations. For the first measurement setup, we'll assume that the two-ships have full communication, but that limited information is passed between the two-ships. Thus, for both TDOA and FDOA, we'll have only two sensor pairs ($\zeta_{2,1}$ and $\zeta_{4,3}$). The second case will include comparison of one sensor from each two-ship ($\zeta_{3,2}$) to complete the nonredundant set. Both of those will be compared against the full measurement set.

We assume 100 ns timing error for TDOA, and 100 Hz frequency error for FDOA, with a carrier frequency of 1 GHz.

Solution: We begin by defining, and plotting, the sensor positions. The geometry is plotted in Figure 3.9.

Next, we define the covariance matrix. For simplicity, we'll assume independence between the TDOA and FDOA measurements. Thus, we define the covariance matrices (in terms of range error for TDOA and range-rate error for FDOA), using the `blkdiag` command to stack the block-diagonal matrices.

```
lam = 3e8/1e9;
time_err = 100e-9; % 100 ns
freq_err = 100; % 10 Hz
rng_err = utils.constants.c*time_err;
rng_rate_err = lam*freq_err;
cov_tdoa = rng_err^2 * eye(n_sensors);
cov_fdoa = rng_rate_err^2 * eye(n_sensors);
cov_full = blkdiag(cov_tdoa, cov_fdoa);
```

Next, we define the different sensor pairs. For ease of coding later, we'll store them in a cell array; the first entry pairs sensors 1 and 2, and sensors 3 and 4. The second also

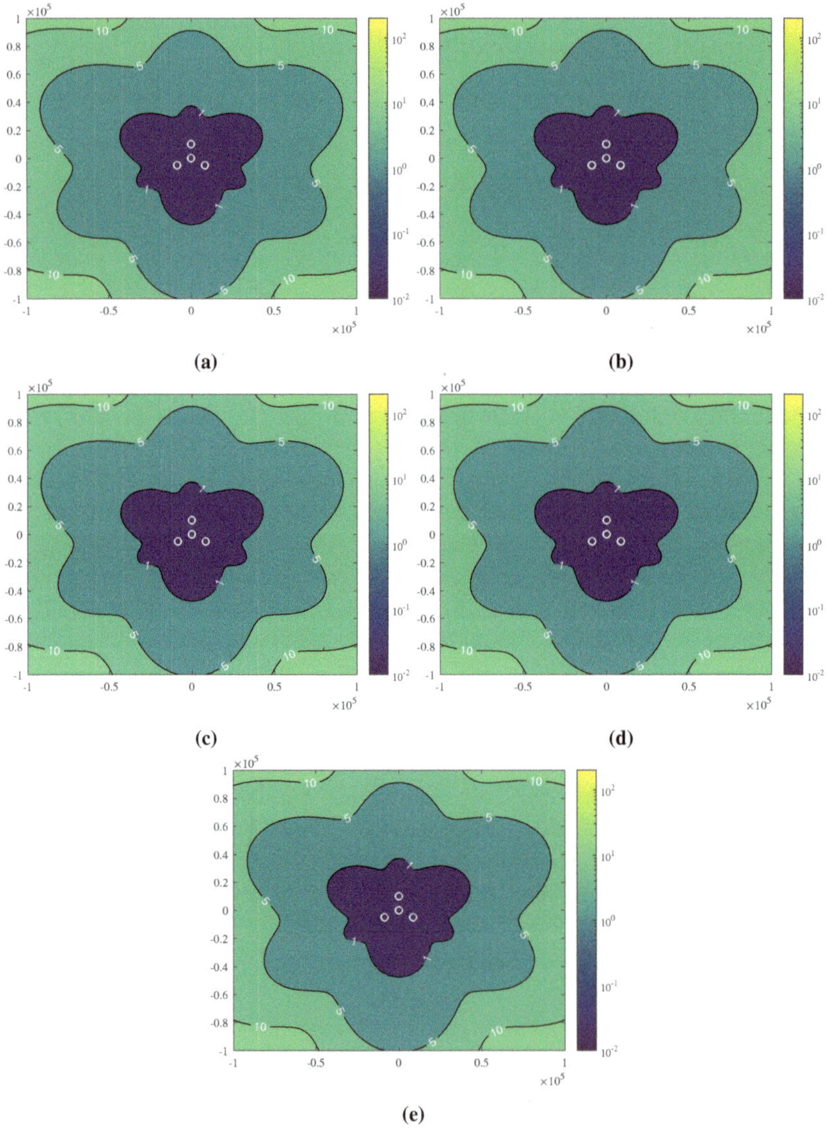

Figure 3.7 TDOA geolocation CEP50 [km] for Example 3.2 with different sensor pairs, based on the CRLB. (a) Sensor 1 as common reference, (b) sensor 2 as common reference, (c) sensor 3 as common reference, (d) sensor 4 as common reference, and (e) full measurement set.

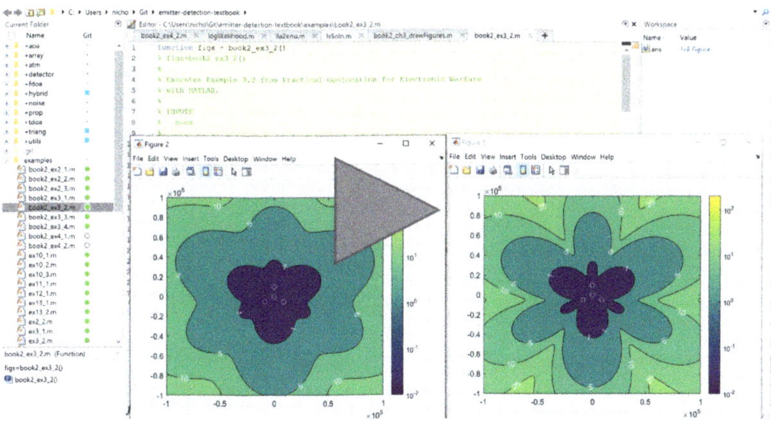

Figure 3.8 Video discussion of 3-D TDOA solution.

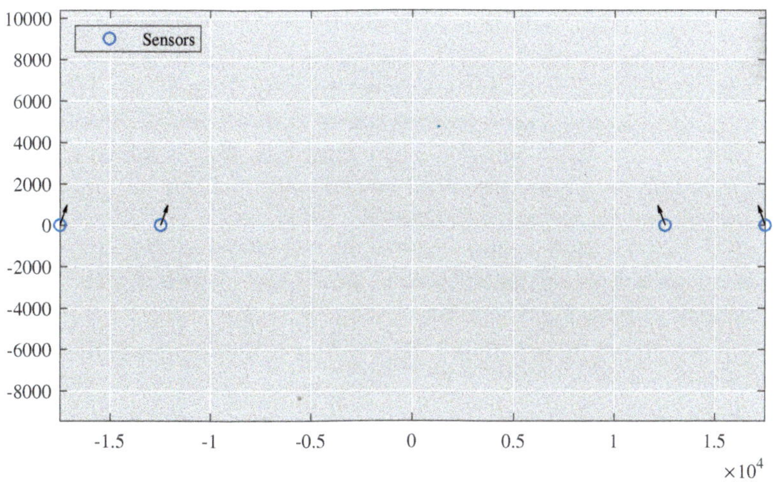

Figure 3.9 Geometry for Example 3.3.

adds a measurement for sensors 2 and 3. Finally, we use the `'full'` shorthand keyword for the full measurement set.

```
ref_set = {[1 3;2 4], [1 2 3;2 3 4], 'full'};
```

Finally, we execute our for loop across the different sensor pair options. With each option, we'll call the `computeCRLB` function to compute the CRLB at each of the test points stored in `x_source`, and then compute the CEP50 for each test point.

```
for idx_set = 1:numel(ref_set)
    this_ref = ref_set{idx_set};

    % Repeat x_sensors for both the TDOA and FDOA sensor sets, because
    % all sensors do both.  Repeat the ref index, because the sensor
    % pairs for TDOA and FDOA are the same.
    this_crlb = hybrid.computeCRLB([],x_sensor, x_sensor, v_sensor, ...
                                   x_source, cov_full, this_ref, ...
                                   this_ref);

    this_rmse = utils.computeCEP50(this_crlb);

    % Plot this result
    ...
end
```

The results are plotted in Figure 3.10 for all three sensor pair configurations. In this case, we see a significant performance increase for the latter two with respect to the first. The former is an incomplete set, and computed only the TDOA and FDOA within each sensor pair (1 against 2, and 3 against 4), but does not compute any TDOA or FDOA estimates across the pairs. This is best considered an *incoherent* combination of the two sensor pairs. This models a situation where intraflight there is a high-speed datalink, but that interflight there is a lower speed datalink, such that only the resultant estimates can be passed, and not the raw data necessary for precision TDOA/FDOA estimation. Inclusion of TDOA/FDOA estimates across the two two-ships adds at least one sensor pair with a significantly increased baseline (25-35 km, instead of 5 km), which leads to a significant increase in geolocation accuracy, as seen in Figures 3.10(b-c). Between the latter two, there is no discernible increase in performance, suggesting that the full measurement set does not improve performance, once a complete nonredundant set is achieved. □

Example 3.4 Multielement TDOA Simulation

In this scenario, we consider a set of five TDOA sensors, to increase the disparity between the number of measurements in the nonredundant (4 measurements) and full (10 measurements) set. A possible justification for this many TDOA sensors is a swarm

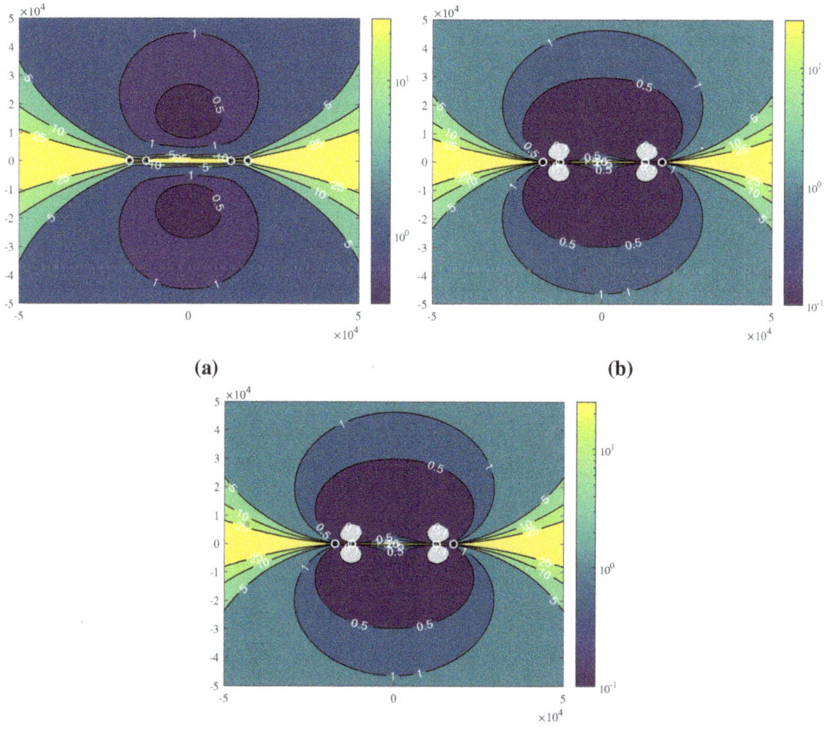

Figure 3.10 TDOA/FDOA geolocation accuracy (CEP50 [km]) for Example 3.3 with different sensor pair configurations, based on the CRLB. (a) Local pairs only (1–2 and 3–4), (b) nonredundant set (1-2, 2-3, 3-4), and (c) full measurement set.

of low-cost drones, such as quad-copters. We will compare performance of classical solvers (maximum likelihood, iterative least squares, and gradient descent) for two sensor selections: a common reference, and the full measurement set.

The true source for this scenario is located at a random location within 100 meters of (3, 4) km, while the sensors are at (1, 0), (3, 0.5), (4, 0), (5, 0.5), and (2, −1) km. The timing accuracy for each sensor's measurement is 100 ns. We will plot the CEP50 as a function of iteration number, averaged over 1,000 Monte Carlo trials, and use a sample solution to show the approach of the iterative solvers. In both plots, we'll compare to the CEP50 computed from the CRLB.

Solution: The sensor coordinates (`x_tdoa`) and source position (`x_source`) are used to generate TDOA measurements for both scenarios. Note that the function `tdoa.measurement` outputs *range*-difference measurements rather than *time*-difference.

```
z_cmn = tdoa.measurement(x_tdoa, x_source);
z_full = tdoa.measurement(x_tdoa, x_source, 'full');
```

Next, we generate the sample covariance matrix for each sensor, in terms of both time of arrival (s^2) and range of arrival (m^2), the former will be called for by `computeCRLB`, but the latter will be used to generate noise to add to the ideal measurements in `z_cmn` and `z_full`. Then, the sample covariance matrix, which is N × N, will be resampled into two measurement covariance matrices, one for the common reference set and one for the full measurement set. Finally, a common set of noise is similarly resamples for each case.

```
% Sample Error Covariance Matrix
err_time = 1e-7; % 100 ns timing error
cov_r = err_time^2 * eye(n_tdoa);
noise = chol(cov_r, 'lower') * randn(n_tdoa, num_mc);

% Resample Covariance Matrices
[test_tdoa,ref_tdoa] = utils.parseReferenceSensor(1,n_tdoa);
cov_z = utils.resampleCovMtx(cov_r, test_tdoa, ref_tdoa);
cov_z = utils.ensureInvertible(cov_z);
noise_z = utils.resampleNoise(noise, test_tdoa, ref_tdoa);

[test_tdoa,ref_tdoa] = utils.parseReferenceSensor('full',n_tdoa);
cov_z_full = utils.resampleCovMtx(cov_r, test_tdoa, ref_tdoa);
cov_z_full = utils.ensureInvertible(cov_z_full);
noise_z_full = utils.resampleNoise(noise, test_tdoa, ref_tdoa);

% Noisy Measurements
zeta = z_cmn + noise_z;
zeta_full = z_full + noise_z_full;
```

The final step before execution of the Monte Carlo trial is to initialize the search parameters for ML, LS, and GD solvers.

```
% ML Search Parameters
x_ctr = [2.5; 2.5]*1e3;
grid_size = [5e3; 5e3];
grid_res = 20;  % meters, grid resolution

% GD and LS Search Parameters
x_init = [1; 1]*1e3;
epsilon = grid_res;
max_num_iterations = 100;
force_full_calc = true;
plot_progress = false;
```

Next, we loop over the noisy measurements, call all three solvers twice (once for each measurement set), and compute the RMSE.

```
for idx=1:num_mc
    % ML Soln
    x_ml = tdoa.mlSoln(x_tdoa, zeta(:,idx), cov_r, x_ctr, grid_size, epsilon
        );
    x_ml_full = tdoa.mlSoln(x_tdoa, zeta_full(:,idx), cov_r, x_ctr,
        grid_size, epsilon,'full');

    % GD Soln
    [x_gd, x_gd_iters] = tdoa.gdSoln(x_tdoa, zeta(:,idx), cov_r, x_init, [],
         [], epsilon, max_num_iterations, force_full_calc, plot_progress);
    [x_gd_full, x_gd_full_iters] = tdoa.gdSoln(x_tdoa, zeta_full(:,idx),
        cov_r, x_init, [], [], epsilon, max_num_iterations, force_full_calc
        , plot_progress, 'full');

    % LS Soln
    [x_ls, x_ls_iters] = tdoa.lsSoln(x_tdoa, zeta(:,idx), cov_r, x_init,
        epsilon, max_num_iterations, force_full_calc, plot_progress);
    [x_ls_full, x_ls_full_iters] = tdoa.lsSoln(x_tdoa, zeta_full(:,idx),
        cov_r, x_init, epsilon, max_num_iterations, force_full_calc,
        plot_progress, 'full');

    % Compute RMSE for this Iteration
    rmse_ml(idx) = norm(x_ml-x_source);
    rmse_gd(idx,:) = sqrt(sum(abs(x_gd_iters-x_source).^2,1));
    rmse_ls(idx,:) = sqrt(sum(abs(x_ls_iters-x_source).^2,1));

    rmse_ml_full(idx) = norm(x_ml_full-x_source);
    rmse_gd_full(idx,:) = sqrt(sum(abs(x_gd_full_iters-x_source).^2,1));
    rmse_ls_full(idx,:) = sqrt(sum(abs(x_ls_full_iters-x_source).^2,1));
```

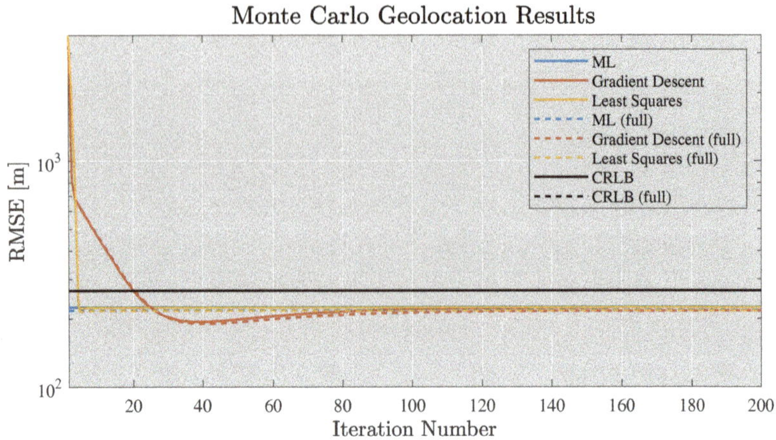

Figure 3.11 Monte Carlo trial RMSE as a function of iteration number for Example 3.4.

```
end
```

Finally, all that is left is to take the average of each set of RMSE calculations, and plot them against the CRLB (for simplicity, we omit the computation of the CRLB here; it follows the same form as in the previous two examples). The RMSE for all solvers is shown as a function of iteration number in Figure 3.11. We see a very similar performance in all cases, with the result that the full measurement set performs slightly better than the nonredundant set, but both are within a few percent of the CRLB-derived estimate. In the example output, shown in Figure 3.12, we see a close-up of the target area, with the 50% error ellipse drawn around the target position, and the solutions for both nonredundant and full measurement sets. Both sets of measurements converge to a solution very near to the target in this case (although it would not be surprising to see either set of measurements converge slightly outside the 50% error ellipse). □

This example has demonstrated how to generate the correlated data samples necessary to test arbitrary sensor selections, and to compare them against the predicted performance via the CRLB.

3.5 PROBLEM SET

3.1 How many TDOA and FDOA measurements are there in the full set of time and frequency measurements from 7 sensors? How many are in the nonredundant set?

3.2 Consider a five-sensor FDOA system with measurement errors of [3, 2, 7, 4, 1] Hz, and a carrier frequency of 3 GHz. Generate the covariance matrix for FDOA measurements in units of $(m/s)^2$ using the second sensor as a common reference.

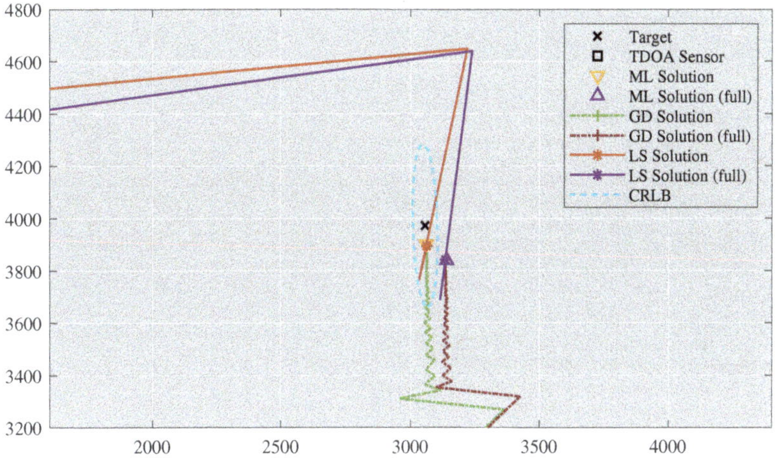

Figure 3.12 TDOA solution geometry for Example 3.4.

3.3 Generate 1,000 measurements using the covariance matrix from Problem 3.2. Use the `cov` command to verify that they have the desired covariance matrix.

3.4 Consider a four-sensor TDOA system with measurement errors of [100, 50, 70, 30] ns. Generate the covariance matrix in units of m^2 using the third sensor as a common reference.

3.5 Generate 1,000 measurements using the covariance matrix from Problem 3.4. Use the `cov` command to verify that they have the desired covariance matrix.

3.6 Generate the covariance matrix for a hybrid geolocation system using the FDOA sensors from Problem 3.2, the TDOA sensors from Problem 3.4, and an AOA sensor with 10° angle error (azimuth-only).

3.7 Consider Example 3.2. Generate noisy measurements for a target at (100, 125) km with each of the five sensor pair configurations, repeat for 100 trials, and compare the results.

3.8 Repeat Example 3.4 with joint TDOA/FDOA sensors that also have a frequency error of 100 Hz and a transmitter carrier frequency of 1 GHz.

References

[1] J. Abel and J. Smith, "The spherical interpolation method for closed-form passive source localization using range difference measurements," *Proc. IEEE International Conference on Acoustics, Speech, and Signal Processing (ICASSP '87)*, vol. 12, pp. 471–474, 1987.

[2] J. Smith and J. Abel, "Closed-form least-squares source location estimation from range-difference measurements," *IEEE Transactions on Acoustics, Speech, and Signal Processing*, vol. 35, no. 12, pp. 1661–1669, 1987. DOI: 10.1109/TASSP.1987.1165089.

[3] J. Abel, "A divide and conquer approach to least-squares estimation," *IEEE Transactions on Aerospace and Electronic Systems*, vol. 26, no. 2, pp. 423–427, 1990. DOI: 10.1109/7.53453.

[4] Y.-T. Chan and K. Ho, "A simple and efficient estimator for hyperbolic location," *IEEE Transactions on Signal Processing*, vol. 42, no. 8, pp. 1905–1915, 1994.

[5] R. O. Schmidt, "A new approach to geometry of range difference location," *IEEE Transactions on Aerospace and Electronic Systems*, vol. AES-8, no. 6, pp. 821–835, 1972. DOI: 10.1109/TAES.1972.309614.

[6] Y. Huang, J. Benesty, G. Elko, and R. Mersereati, "Real-time passive source localization: A practical linear-correction least-squares approach," *IEEE Transactions on Speech and Audio Processing*, vol. 9, no. 8, pp. 943–956, 2001. DOI: 10.1109/89.966097.

[7] J. F. Reale, L. M. Huie, and M. L. Fowler, "Locating closely spaced coherent emitters using TDOA techniques," in *2013 Asilomar Conference on Signals, Systems and Computers*, 2013, pp. 2060–2064. DOI: 10.1109/ACSSC.2013.6810691.

[8] W. H. FOY, "Position-location solutions by taylor-series estimation," *IEEE Transactions on Aerospace and Electronic Systems*, vol. AES-12, no. 2, pp. 187–194, 1976. DOI: 10.1109/TAES.1976.308294.

[9] K. Ho and Y. Chan, "Solution and performance analysis of geolocation by tdoa," *IEEE Transactions on Aerospace and Electronic Systems*, vol. 29, no. 4, pp. 1311–1322, 1993. DOI: 10.1109/7.259534.

[10] K. C. Ho and W. Xu, "An accurate algebraic solution for moving source location using TDOA and FDOA measurements," *IEEE Transactions on Signal Processing*, vol. 52, no. 9, pp. 2453–2463, Sep. 2004, ISSN: 1053-587X. DOI: 10.1109/TSP.2004.831921.

[11] R. Kaune, "Accuracy studies for TDOA and TOA localization," in *2012 15th International Conference on Information Fusion*, Jul. 2012, pp. 408–415. [Online]. Available: http://fusion.isif.org/proceedings/fusion12CD/html/pdf/056_271.pdf.

[12] B. Yang and J. Scheuing, "A Theoretical Analysis of 2D Sensor Arrays for TDOA Based Localization," in *2006 IEEE International Conference on Acoustics, Speech, and Signal Processing Proceedings*, vol. 4, 2006. DOI: 10.1109/ICASSP.2006.1661115.

[13] B. Yang and J. Scheuing, "Cramer-rao bound and optimum sensor array for source localization from time differences of arrival," in *Proceedings.(ICASSP'05). IEEE International Conference on Acoustics, Speech, and Signal Processing, 2005.*, IEEE, vol. 4, 2005, pp. iv–961.

[14] Y. Wang and K. Ho, "TDOA positioning irrespective of source range," *IEEE Transactions on Signal Processing*, vol. 65, no. 6, pp. 1447–1460, 2017.

[15] Y. Wang and K. C. Ho, "Unified near-field and far-field localization for AOA and hybrid AOA-TDOA positionings," *IEEE Transactions on Wireless Communications*, vol. 17, no. 2, pp. 1242–1254, Feb. 2018, ISSN: 1536-1276. DOI: 10.1109/TWC.2017.2777457.

[16] M. Pourhomayoun and M. L. Fowler, "Distributed computation for direct position determination emitter location," *IEEE Transactions on Aerospace and Electronic Systems*, vol. 50, no. 4, pp. 2878–2889, 2014. DOI: 10.1109/TAES.2014.130005.

[17] M. Pourhomayoun, M. L. Fowler, and N. E. Wu, "Spatial sparsity based emitter localization," in *2012 46th Annual Conference on Information Sciences and Systems (CISS)*, 2012, pp. 1–4. DOI: 10.1109/CISS.2012.6310710.

[18] H. C. So, Y. T. Chan, and F. K. W. Chan, "Closed-form formulae for time-difference-of-arrival estimation," *IEEE Transactions on Signal Processing*, vol. 56, no. 6, pp. 2614–2620, 2008. DOI: 10.1109/TSP.2007.914342.

[19] R. Ren, M. L. Fowler, and N. E. Wu, "Finding optimal trajectory points for TDOA/FDOA geo-location sensors," in *2009 43rd Annual Conference on Information Sciences and Systems*, 2009, pp. 817–822. DOI: 10.1109/CISS.2009.5054830.

[20] X. Hu and M. L. Fowler, "Sensor selection for multiple sensor emitter location systems," in *2008 IEEE Aerospace Conference*, 2008, pp. 1–10. DOI: 10.1109/AERO.2008.4526431.

[21] J. Velasco, D. Pizarro, J. Macias-Guarasa, and A. Asaei, "Tdoa matrices: Algebraic properties and their application to robust denoising with missing data," *IEEE Transactions on Signal Processing*, vol. 64, no. 20, pp. 5242–5254, 2016. DOI: 10.1109/TSP.2016.2593690.

[22] L. L. Scharf, *Statistical Signal Processing*. Reading, MA: Addison-Wesley, 1991.

[23] N. A. O'Donoughue, *Emitter Detection and Geolocation for Electronic Warfare*. Norwood, MA: Artech House, 2019.

[24] B. Carlson, "Covariance matrix estimation errors and diagonal loading in adaptive arrays," *IEEE Transactions on Aerospace and Electronic Systems*, vol. 24, no. 4, pp. 397–401, 1988. DOI: 10.1109/7.7181.

Chapter 4

Coordinate Systems

The software and algorithms that are shown in this textbook, and indeed in the majority of the research literature, assume a Cartesian coordinate system, such as the one shown in Figure 4.1(a). This is convenient from a mathematical perspective, as Cartesian geometry is well expressed in linear algebra. The world, however, is not conducive to Cartesian reference frames. Targets are often more naturally expressed in spherical coordinates (bearing - θ, elevation - ϕ, and range - r), shown in Figure 4.1(b), and global coordinates are frequently geodetic (latitude - ϕ, longitude - λ, and altitude – or height above the Earth's surface - h), shown in Figure 4.1(c).

In this chapter, we'll briefly discuss several common coordinates systems, and how to convert them to a suitable Cartesian coordinate system, so that the utilities and algorithms presented can be leveraged. We will first discuss *local* coordinate systems in Section 4.1, then introduce models of the Earth and global coordinate systems in Section 4.2. Section 4.3 discusses the usage of the provided tools for converting between common coordinate systems, including a few example problems.

4.1 LOCAL COORDINATE SYSTEMS

In this section, we'll discuss local coordinate systems, which represent positions relative to a reference point, such as a point of interest or a sensor's position.

4.1.1 Cartesian Coordinates

Cartesian coordinates, as discussed earlier, consist of three axes, typically labeled x, y, and z. The three axes are all perpendicular to each other, and are oriented such that the *cross-product* of x and y pointing vectors results in the z axis' pointing vector. Any of the algorithms we have derived or provided in the accompanying MATLAB® code will work natively with any local Cartesian coordinate system.

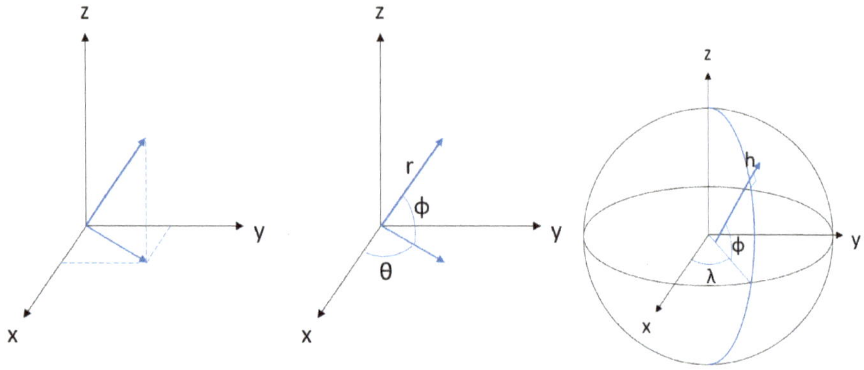

Figure 4.1 Example coordinate systems: (a) Cartesian, (b) spherical, and (c) geodetic.

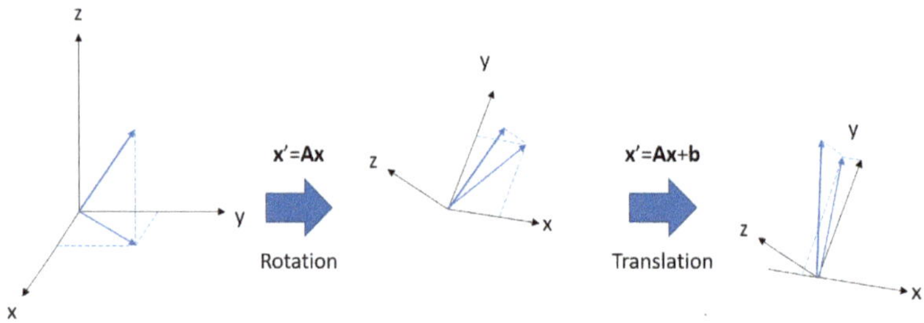

Figure 4.2 Illustration of the two primary operations of an affine transformation, used to translate Cartesian coordinates from one reference frame to another.

Cartesian coordinates provide a link between Euclidean geometry and algebra, allowing for convenient expression and manipulation of coordinates with linear algebra, and simple evaluation of quantities such as the Euclidean norm (distance) [1]. Computation of the pointing vector from one point to another is as simple as subtracting its coordinates.

Conversion between one Cartesian coordinate system and another is an affine transformation, parameterized by a rotation matrix $\mathbf{A}$, to align the x, y, and z axes in the two coordinate systems, and a translation vector $\mathbf{b}$ to shift the origin. This process is illustrated in Figure 4.2. While the point being referenced does not move, the reference frame is rotated and translated around it.

$$\mathbf{x}' = \mathbf{A}\mathbf{x} + \mathbf{b}. \tag{4.1}$$

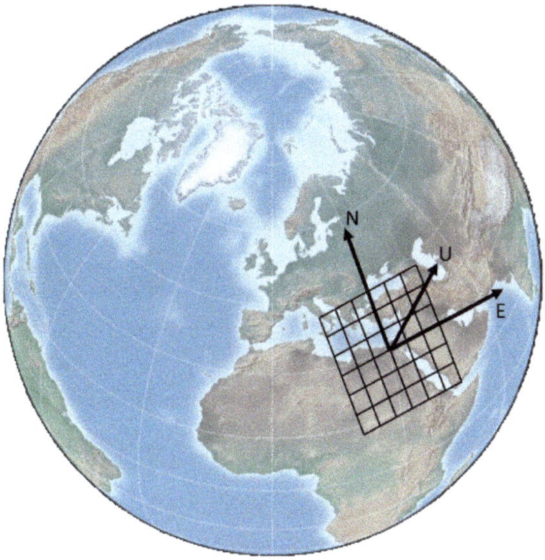

Figure 4.3 Illustration of the East, North, Up (ENU) coordinate system, Earth imagery from [2].

4.1.1.1 East, North, Up (ENU) Coordinates

East, North, Up (ENU) is a specific Cartesian coordinate system, and is a hybrid between local and global reference frames. The origin and rotation of an ENU coordinate system depend on the reference coordinate (ϕ_0, λ_0).[1] The origin is a place along the Earth's surface (altitude = 0 m)[2] at those coordinates, and the axes are oriented such that the x- and y-axes are parallel to the Earth's surface at that point, with x pointing east and y pointing north. The z-axis, then, points up (perpendicular to the local ground plane). This is shown in Figure 4.3.

4.1.2 Spherical Coordinates

Spherical coordinates represent a point in three-dimensional space with a distance (the Euclidean norm of the vector from the origin to that point), as well as two angles: the angle between the pointing vector and the horizontal plane (elevation), as well as the

1. As noted earlier in this chapter, latitude is commonly expressed with the variable ϕ while longitude is expressed with λ. We will use those terms in this chapter, even though λ is elsewhere used to refer to a signal's wavelength and ϕ is often the elevation angle in local spherical coordinates. We will attempt to be clear in our usage of these ambiguous terms, but encourage the reader to be vigilant when reading to avoid confusion.
2. The reference point can be placed at any altitude above the Earth's surface, but it is a best practice to leave this at the surface of the Earth to avoid confusion.

angle within the horizontal plan (bearing or azimuth). The azimuth angle is typically defined with the $+x$ axis as its origin (0 degrees or radians) and the $+y$ axis as the positive direction (counterclockwise), but is occasionally defined with $+y$ as its origin, and $+x$ as a positive angle (clockwise).[3.]

These values are computed from a Cartesian coordinate $\mathbf{x} = (x, y, z)$ with:

$$r = \sqrt{x^2 + y^2 + z^2} = \|\mathbf{x}\|_2 \tag{4.2}$$

$$\phi = \sin^{-1}\left(z/\sqrt{x^2 + y^2}\right) \tag{4.3}$$

$$\theta = \sin^{-1}(y/x) \tag{4.4}$$

where the azimuth θ is here defined counterclockwise from $+x$. The reverse transformation is computed

$$x = r\cos(\phi)\cos(\theta) \tag{4.5}$$

$$y = r\cos(\phi)\sin(\theta) \tag{4.6}$$

$$z = r\sin(\phi) \tag{4.7}$$

This conversion is implemented in the built-in MATLAB® tools `sph2cart` and `cart2sph`, which assume the convention above, that θ is defined CCW from +x. Local spherical coordinates are sometimes referred to as *azimuth, elevation, and range* (AER), which we do in this text. We will use the convention for AER that azimuth is defined in the reverse (CW from +y). Since this is different from MATLAB®'s standard convention, we have implemented some conversion tools, namely `utils.enu2aer` and `utils.aer2enu` that handle the conversion between spherical (AER) and Cartesian (ENU) coordinates.[4.]

The most important observation here is that, while the transformation is very simple, it is nonlinear. And, so, it can be computationally complex to translate a lot of coordinates from one dimension to another, or to continually operate in spherical coordinates. For example, expression of the pointing vector from one point to another in spherical coordinates requires translation of both spherical coordinates to Cartesian coordinates, calculation of the pointing vector in Cartesian space (via subtraction of the two coordinates), and then conversion back to spherical coordinates using the formulas above. Thus, continual operation in spherical coordinates will be much more computationally intensive, since every operation or comparison of points requires this 3-step process. Often, the best approach is to work in Cartesian space, and then handle conversion of coordinates at the input and output stages alone.

3. This is a common source of confusion or errors. For this reason, it is wise to be clear about which convention is in use whenever recording or transmitting spherical coordinates.
4. In our conversion, we assume that AER is oriented to align the $+x$ axis to east and $+y$ axis to north.

4.2 GLOBAL COORDINATE SYSTEMS

When applying geolocation approaches to the real world, it can be helpful to express positions using global coordinates, particularly as the standoff ranges grow larger. *Geodetic* coordinates (degrees latitude and longitude, along with altitude above the Earth's surface) are universally understood, and are often the preferred input and output coordinate format, but it is not the only global coordinate system. A second popular system, *Earth-centered Earth-fixed* (ECEF) is also global, but is a Cartesian coordinate system, with all of the computational and algorithmic advantages of local Cartesian coordinates [3]. So, it is often the best approach to convert from geodetic coordinates into ECEF for processing, and then back to geodetic for output.

The easiest way to work with geodetic coordinates in MATLAB® is via the *Mapping Toolbox*, which has tools available for conversion between these coordinate systems, and others, as well as for the analysis and display of geospatial data. In this code suite, however, we have implemented a subset of the geospatial algorithms that may be useful for those without access to the *Mapping Toolbox*.

A third system, *Earth-centered inertial* (ECI), is defined similarly to ECEF, but is fixed with respect to the stars, rather than the Earth, and so it is more convenient for description of celestial bodies, such as satellites [4]. We will briefly discuss this coordinate system, but do not provide any tools for it, and instead refer users to MATLAB®'s *Aerospace Toolbox*, which can provide the necessary tools.

4.2.1 Models of the Earth

Several models of the Earth, with varying levels of complexity and accuracy, are illustrated in Figure 4.4.

4.2.1.1 Flat Earth

The first, and simplest, is a flat-Earth approximation. Here, we assume that there is no curvature, and rely on a local reference frame, such as ENU. This is convenient for short range, but is problematic for distances greater than 20 km, and has significant errors beyond 100 km due to the Earth's curvature. After 20 km, the ground reference is off by 31 meters, and at 100 km it is off by almost 785 m.[5] This is particularly problematic for points near the horizon, so it is often recommended not to use a flat-Earth approximation for low-altitude targets or those at long-range.

Nevertheless, for high-altitude targets, where occlusion from the horizon is not a concern, a flat-Earth model presents a convenient reference frame for efficient computation, and aligns well with ENU coordinates (although at long-range there is a significant difference between a target's local altitude and it's "up" coordinate, which is not captured by a flat-Earth model).

5. This is based on an approximate radius of 6,400 km [5].

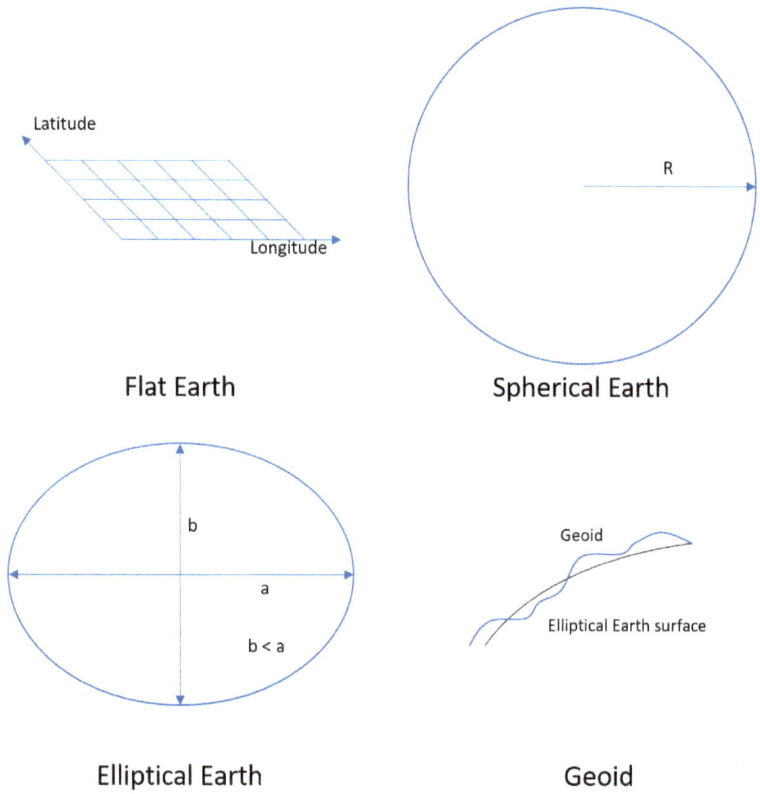

Figure 4.4 Models of the Earth.

4.2.1.2 Spherical Earth

The second approximation is a spherical-Earth model, which greatly increases fidelity, particularly for modeling of the horizon, and can often provide sufficient accuracy for performance analysis and modeling, but it is not suitable for navigation. If extreme accuracy is not required, this should be sufficient. The recommended radius is 6,400 km for back of the envelope calculations, while the International Union of Geodesy and Geophysics (IUGG) recommends 6,371 km [5], which varies from the true radius at any point on the Earth's surface by as much as 0.3%. With this radius, the maximum error in calculation of distance between two points on the globe is on the order of 1%.

If you have the mapping toolbox, then the command `earthRadius` will return a spherical average radius of the Earth; otherwise use the variable `Re_true` in the `utils.constants` structure.

When modeling the propagation of radio waves, we must consider the effects of refraction, which is the curvature of waves as they pass through a changing medium, such as the Earth's atmosphere. The effects of refraction can often be ignored, particularly for vertical paths or at high altitude, but for long-range and low-altitude links, the effects of refraction can be conveniently approximated by modeling the radius of the Earth with a multiplier of 4/3 [6], which can be referenced with `utils.constants.Re`. The net result of this approximation is that the radio horizon occurs at a slightly longer range than a straight line would suggest, and that incoming signals arrive at a slightly higher elevation angle than a straight line would predict.

4.2.1.3 Elliptical Earth

The Earth's shape is more closely approximated with an ellipsoid than a sphere, as its rotation causes a slight flattening; the radius at the equator (6,378.137 km) is slightly larger than the radius measure through both poles (6,356.137 km) [7]. Two elliptical models of the Earth are the GRS-80 model [5] and the WGS-84 model [7]. Significant parameters are summarized in Table 4.1; note that they are almost identical.[6] MATLAB®'s Mapping Toolbox has an implementation of WGS84 that can be accessed via the `wgs84ellipsoid` function, but we have provided some of these parameters in `utils.constants` in the variables `semimajor_axis_km`, `semiminor_axis_km`, `first_ecc_sq`, and `second_ecc_sq`.[7]

4.2.1.4 Geoid Earth

The Earth is not homogeneous, and so there are slight fluctuations in the gravitational field across the globe. This is of minor importance to the geolocation of terrestrial

6. WGS-84 is the most commonly used model for an elliptical Earth. Most GPS equipment, for example, uses WGS-84 as its reference ellipsoid, and most terrain data is referenced to the WGS-84 ellipsoid [8].
7. Note that the WGS-84 model has been updated multiple times, most recently in 2021 [9].

Table 4.1
Ellipsoid Model Parameters

	GRS-80	WGS-84
Semi-major axis length [km]	6,378.137	6,378.137
Semi-minor axis length [km]	6,356.752 314 140	6,356.752 314 245
First eccentricity squared	6.694 380 022 90e-3	6.694 379 990 14e-3
Second eccentricity squared	6.739 496 775 48e-3	6.739 496 742 28e-3

Data from [5], [7].

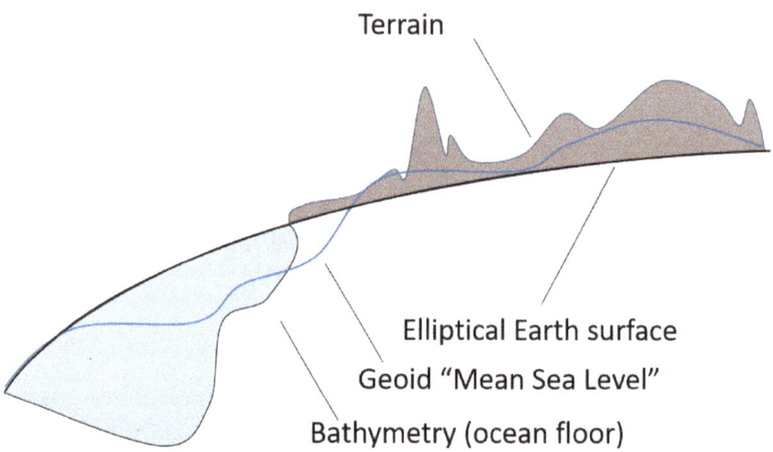

Figure 4.5 Illustration of elliptical and geoid Earth models and terrain. Drawn with inspiration from [10].

targets, but affects orbital calculations. As such, a geoid model can be used to represent surfaces of constant gravitational potential. For Earth, the GRM90 model expresses what the mean sea level (MSL) would be across the Earth's surface, in the absence of land, waves, and tidal forces. It is expressed as a geoid height above (or below) the WGS-84 elliptical model. The relationship between the WGS84 elliptical model, the GRM90 geoid model, and terrain data is shown in Figure 4.5 (albeit a very exaggerated illustration).

We don't include any routines to work with geoid models of the Earth, since they do not significantly affect our algorithms.

4.2.1.5 Terrain

For increased accuracy, particularly for propagation modeling or line of sight blockage calculations, terrain must be considered. Terrain data is available for much of the world at varying resolutions. MATLAB® has built-in tools to fetch DTED and other geospatial data (like imagery, roads, borders, lakes, and population centers) from various online resources, such as EarthExplorer (maintained by U.S. Geological Survey) [11], and Data.gov (maintained by U.S. General Services Administration) [12]. Users with Mapping Toolbox can use the `wmsfind` and `wmsread` functions to obtain the data.

The most comprehensive set of terrain data that is publicly available was collected by the Shuttle Radio Tomography Mission (SRTM), which is provided at two resolutions: SRTM1 (1 arc-second, approximately 30 m resolution), and SRTM3 (3 arc-seconds, approximately 90 m resolution), for all points between 60°S and 60°N [13].

Similar to geoid models, terrain data is expressed as the terrain height above an ellipsoidal Earth (WGS-84), as illustrated in Figure 4.5.

The most significant drawback to using terrain data is that it ties any analysis or results to a specific location. This is suitable, even desirable, when planning or assessing specific scenarios, but is often a detriment for system design or algorithm performance assessment as it prevents the generalization of results to other locations. For this reason, we do not explicitly consider the effects of terrain in any of the provided algorithms in this text.

4.2.2 Latitude and Longitude

The global coordinate system most widely taught and commonly understood is geodetic, referred to as latitude and longitude or, more completely, latitude, longitude, and altitude (LLA). Lines of constant latitude are parallel rings around the Earth that are perpendicular to its axis of rotation, spanning 90° north and south, with the equator at 0°. Lines of longitude are perpendicular to every line of latitude, and run along the Earth's surface. Due to the curvature of the Earth, they are not parallel with each other, rather they all intersect at the north and south poles. Altitude is perpendicular to the local ground plane at a given Lat/Lon coordinate. This is shown in Figure 4.6.

Since they are parallel, the distance between two lines of latitude is constant (approximately 111 km per degree of latitude). The distance between degrees of longitude is, however, latitude-dependent. At the equator, it is approximately 111 km per degree of longitude, while at 40 degrees north or south, it is approximately 85 km per degree of longitude [14].

It is very convenient to express points on or near the Earth's surface in terms of latitude, longitude, and altitude, but because it is not Cartesian, this coordinate system presents significant challenges for calculation of distances, bearing, and motion.

Latitude and longitude are typically expressed in geodetic coordinates, which are shown in Figure 4.6, where latitude is the angle between the line perpendicular to

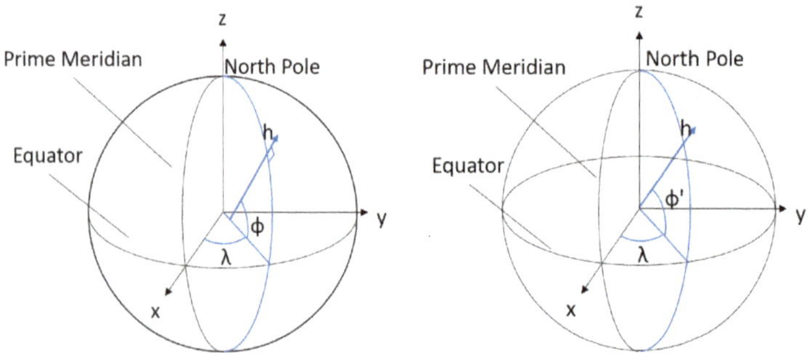

Figure 4.6 Illustration of latitude and longitude using (a) geodetic and (b) geocentric coordinates. Note that for geodetic coordinates the pointing vector for latitude (*phi*) does not intersect with the origin, rather it is defined to be perpendicular to the Earth's surface, while the geocentric pointing vector (for ϕ') does.

the Earth's surface, and the equatorial plane. Geocentric latitude, by contrast, uses the pointing vector from the origin to the surface. In this text, we use geodetic coordinates exclusively for latitude and longitude.

4.2.3 Earth-Centered Earth-Fixed (ECEF)

Earth-centered Earth-fixed is a Cartesian coordinate system, fixed to the Earth, with its origin at the center, its x-axis oriented toward (0°N, 0°E), the point where the prime meridian and the equator intersect, its y-axis oriented to (0°N, 90°E), and its z-axis oriented towards the north pole, as shown in Figure 4.7. Since it is a Cartesian coordinate system, it is usable directly in the provided MATLAB® code; no conversions are necessary.

Since ECEF is fixed to the Earth's center, and to geodetic coordinates, it is an appealing coordinate system in which to work for points on the Earth's surface or in the atmosphere. Objects in orbit can be referenced in ECEF, although their high velocity, and the intricacies of orbital mechanics, require more sophisticated equations. Additionally, the Earth's surface is not easily represented, and so computation of horizon range or terrain blockage is more complex.

4.2.4 Earth-Centered Inertial (ECI)

When discussing celestial objects, such as stars, planets, or satellites, most astronomers and engineers use *inertial* reference frames, such as *Earth-centered inertial* (ECI). Inertial reference frames have a nonaccelerating origin, and do not rotate with respect to the stars. An informative, but short, history on celestial navigation and its relation to ECI

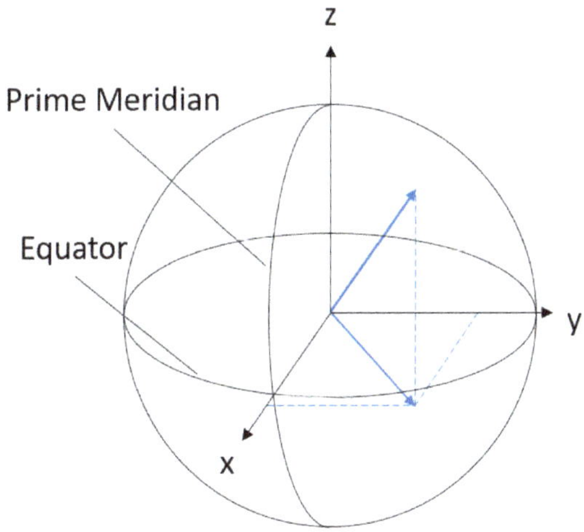

Figure 4.7 Illustration of Earth-centered Earth-fixed (ECEF) coordinate system.

coordinates, is found in [15]. A primary benefit of inertial frames, such as ECI, is that orbiting objects can be compactly expressed as a six-element state vector (x/y/z position and velocity at the reference time) [16].

ECI is a Cartesian reference frame, defined similarly to ECEF, except that it does not rotate with the Earth as ECEF does. Every ECI coordinate system is referenced to the Earth's orientation at a specific point in time, such as J2000 (January 1, 2000 at 12:00:00 AM), although any time point can be used [17].

Conversion from ECI to ECEF or to another rotating reference frame is a complex process that involves corrections for: (1) time since the reference epoch, (2) the difference between rotational and atomic time, and (3) complexities of orbital mechanics. MATLAB® users with the *Aerospace Toolbox* can leverage functions such as `eci2ecef` for conversion to ECEF, `eci2lla` for conversion to LLA, or `eci2aer` to convert to local spherical coordinates.

4.3 USAGE

To assist with usage, we have provided functions that convert between coordinate systems. Conversion between any set of coordinates in AER, ENU, ECEF, or LLA can be achieved with three fundamental functions, shown in Figure 4.8. For convenience, we also include wrappers, so that users can translate between any of these four types without having to understand the process.

The functions are found in the `utils` namespace, and take the form <x>2<y> where <x> and <y> are: `aer`, `ecef`, `enu`, or `lla`. We do not include any conversions

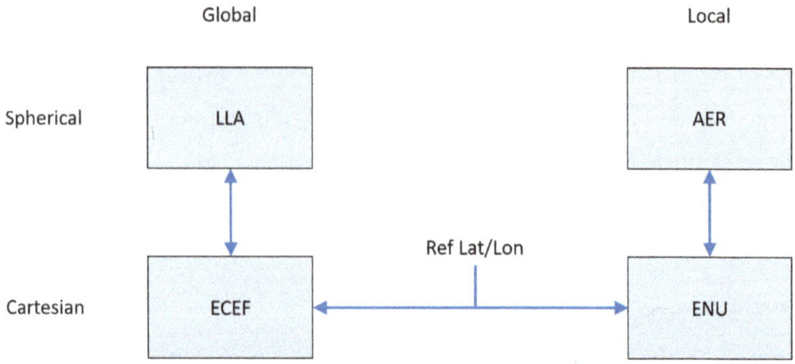

Figure 4.8 Block diagram of implemented conversion functions for various coordinate systems.

for ECI (since that is outside the scope of this book), or for Cartesian to/from spherical coordinates (since that is included in MATLAB®'s base functionality). Across all of the functions, there are optional string inputs for `dist_units` and `angle_units`, the former must be one of the recognized MATLAB® distance types (such as `'m'` or `'km'`) for all of the Cartesian or distance coordinates, while the latter must be either `'rad'` or `'deg'`, indicating what units the spherical and geodetic angles are expressed in.

Conversion between local coordinate systems ENU and AER is discussed in Section 4.1.

4.3.1 Conversion between ECEF and LLA

The most complex conversion is between ECEF and LLA. Many exact approaches and approximations have been proposed over the decades [18]. Many attempt to find the most computationally efficient approach [19], while others trade accuracy for reduced complexity [20], and still more derive iterative estimates [21], [22]. We have implemented the closed-form algorithm presented in [23], which has a few simple steps. It is not the most accurate algorithm, but it is suitable for research purposes.

4.3.1.1 ECEF to LLA

Let the ECEF coordinate be expressed (x, y, z), in meters. To convert to LLA, we first compute some auxiliary parameters, including the length of the pointing vector in the equatorial plane (p), an auxiliary angle term (θ) [23]:

$$p = \sqrt{x^2 + y^2} \tag{4.8}$$

$$\theta = \tan^{-1}\left(\frac{za}{pb}\right) \tag{4.9}$$

where a is the semimajor axis length (in m), and b is the semiminor axis length (in m). Next, we estimate the latitude (ϕ), longitude (λ), and altitude (h) with:

$$\phi = \tan^{-1}\left(\frac{z + e_2^2 b \sin^3(\theta)}{p - e_1^2 a \cos^3(\theta)}\right) \tag{4.10}$$

$$\lambda = \tan^{-1}\left(\frac{y}{x}\right) \tag{4.11}$$

$$h = \frac{p}{\cos(\lambda)} - R_{\text{eff}}(\phi) \tag{4.12}$$

where e_1^2 and e_2^2 are the first and second eccentricity squared, respectively (see Table 4.1), and $R_{\text{eff}}(\phi)$:

$$R_{\text{eff}}(\phi) = \frac{a}{\sqrt{1 - e_1^2 \sin^2(\phi)}} \tag{4.13}$$

This is implemented in `utils.ecef2lla`, and can be used with the command:

```
> [lat, lon, alt] = utils.ecef2lla(x, y, z);
```

4.3.1.2 LLA to ECEF

With the effective radius defined in (4.13), the conversion from LLA to ECEF is a simple trigonometric calculation

$$x = (R_{\text{eff}} + h)\cos(\phi)\cos(\lambda) \tag{4.14}$$

$$y = (R_{\text{eff}} + h)\cos(\phi)\sin(\lambda) \tag{4.15}$$

$$z = \left(\left(1 - e_1^2\right) R_{\text{eff}} + h\right)\sin(\phi) \tag{4.16}$$

This is implemented in `utils.lla2ecef`, and can be used with the command:

```
> [x, y, z] = utils.lla2ecef(lat, lon, alt);
```

4.3.2 Conversion between ENU and ECEF

Both ENU and ECEF are Cartesian coordinate systems, so conversion is a simple affine transformation. In addition to the input coordinates, conversion requires the reference point for the ENU coordinate frame (in degrees latitude/longitude) in order to define the translation and rotation terms.

4.3.2.1 ECEF to ENU

The first step is to convert the reference point from LLA to ECEF coordinates and take the difference. This handles the *translation* portion of the affine transformation, representing the point in local Cartesian coordinates for the reference point but still oriented to ECEF.

```
[x0, y0, z0] = utils.lla2ecef(lat0, lon0, alt0);
```

where `lat0`, `lon0`, and `alt0` are the reference coordinates for the the ENU coordinate frame. The pointing vector from the ENU origin to the point (x, y, z) is computed via subtraction (although still expressed in the orientation of the ECEF coordinate frame).

$$\delta_x = x - x_0 \tag{4.17}$$

$$\delta_y = y - y_0 \tag{4.18}$$

$$\delta_z = z - z_0 \tag{4.19}$$

Then, the rotation is achieved with a few simple equations, noting that t is a dummy variable used to express the size of the pointing vector within the equatorial plane.

$$t = \cos(\lambda_0)\delta_x + \sin(\phi_0)\delta_y \tag{4.20}$$

$$e = -\sin(\lambda_0)\delta_x + \cos(\lambda_0)\delta_y \tag{4.21}$$

$$n = -\sin(\phi_0)t + \cos(\phi_0)\delta_z \tag{4.22}$$

$$u = \cos(\phi_0)t + \sin(\phi_0)dz \tag{4.23}$$

This is implemented in `utils.ecef2enu` and can be used with the command:

```
> [e, n, u] = utils.ecef2enu(x, y, z, lat0, lon0, alt0);
```

4.3.2.2 ENU to ECEF

The inverse transformation takes the same form, but in reverse order. First, we reverse the rotation of coordinate frames from local east, north, up to the ECEF orientation:

$$t = -\sin(\phi_0)n + \cos(\phi_0)u \tag{4.24}$$

$$dx = -\sin(\lambda_0)e + \cos(\lambda_0)t \tag{4.25}$$

$$dy = -\cos(\lambda_0)e + \sin(\lambda_0)t \tag{4.26}$$

$$dz = \cos(\phi_0)n + \sin(\phi_0)u \tag{4.27}$$

The local offsets are now oriented to ECEF, and can be translated by adding the ECEF coordinates of the reference point.

```
[x0, y0, z0] = utils.lla2ecef(lat0, lon0, alt0);
```

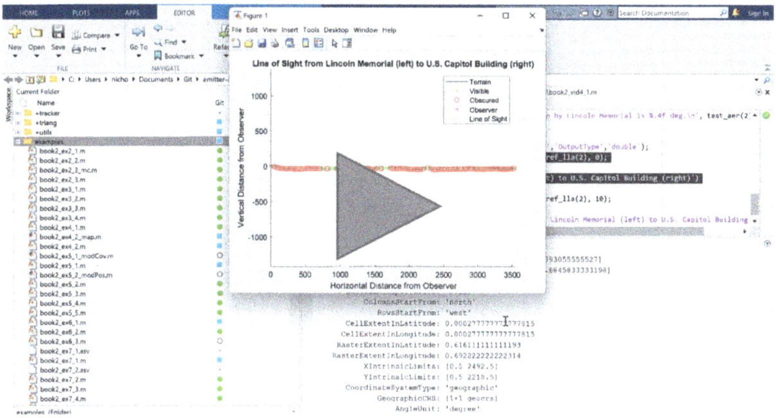

Figure 4.9 Video discussing conversion between geodetic, global and local Cartesian, and local spherical coordinate frames.

$$x = x_0 + \delta_x \tag{4.28}$$

$$y = y_0 + \delta_y \tag{4.29}$$

$$z = z_0 + \delta_z \tag{4.30}$$

This is implemented in `utils.enu2ecef`, and can be used with the command:

```
> [x, y, z] = utils.enu2ecef(e, n, u, lat0, lon0, alt0);
```

The video linked in Figure 4.9 walks through some of the coordinate conversions in this text.

4.3.3 ENU Adapted for Spherical Earth

As noted earlier, the ENU reference frame assumes a flat-Earth, and there will be a difference of a few hundred meters between the *Up* coordinate and local altitude for points more than 100 km from the reference of the ENU coordinates.

To account for this, we have implemented a pair of utility functions, `correctENU` and `reckonSphereENU`. The latter accepts a distance `R_g` that represents a distance traveled along the Earth's surface, and an altitude `alt` for some point. They are converted to an *Up* coordinate, and a distance `R_en` in the east-north plane. The former, `correctENU` accepts a 3-D coordinate, in meters east and north (along the Earth's surface) and a local altitude, then uses `reckonSphereENU` to return the coordinates of that same point in local ENU.

The value of these utilities is primarily in being able to specify points with a distance along the Earth's surface and a local altitude, and to easily convert those into Cartesian ENU for analysis or plotting.

4.3.4 Examples

In this section, we present two examples that illustrate not only how to use the provided converters to handle common coordinate systems, but also how to do so in the context of geolocation problems, and predicting geolocation performance.

Example 4.1 Detection Example Using LLA

Consider three sensors in a coaltitude formation aligned east-west, with a spacing of 5 km, and heading east at 80 km/hr. The central sensor is located at 5°N, 15°W, at an altitude of 10 km. These sensors have AOA accuracy of 2°, in azimuth and elevation, TDOA accuracy of 1 μs, and an FDOA accuracy of 100 Hz against an emitter at 4° 30′ N, 14° 40′ W, and 0 m altitude (located on the Earth's surface) transmitting a signal centered at 1 GHz.

Generate noisy measurements and use iterative least squares to estimate the source position, and report that estimate's position in LLA. Compute the miss distance, report it in meters, and compare it to the CRLB.

Solution: First, we define the sensors using the provided coordinates. The simplest approach is in ENU units:

```
ref_lat = 5;    % deg N
ref_lon = -15; % deg E
ref_alt = 0;    % m
x_sensor_enu = [-5, 0, 5;
                 0, 0, 0;
                 10, 10, 10]*1e3;     % m
kph2mps = 1e3/60^2;                  % kph -> m/s
v_sensor_enu = 80* [1, 1, 1;
                    0, 0, 0;
                    0, 0, 0]*kph2mps; % m/s
source_lat = 4.5;                    % deg N
source_lon = -(14 + 40/60);          % deg E
source_alt = 0;                      % m
v_source_enu = [0, 0, 0];            % m/s
```

Next, to put them in the same coordinate frame, we convert the source position from LLA to ENU

```
[e, n, u] = utils.lla2enu(source_lat, source_lon, source_alt, ref_lat,
    ref_lon, ref_alt);
x_source_enu = [e, n, u];
```

To set up our solution, we generate the clean (z) and noisy (ζ) measurements. This takes some care, as each of the sensors will measure AOA, TOA, and FOA, so we begin by defining our covariance matrices for each measurement type and assembling the complete covariance matrix across all measurements (cov_x), and generate a noise vector.

```
% Define error terms
source_freq_Hz = 1e9;
err_aoa_deg = 2;
err_tdoa_s = 1e-6;
err_fdoa_Hz = 10;

% Compute error covariance (measurement level)
c = utils.constants.c; % speed of light
cov_psi = (err_aoa_deg*pi/180)^2*eye(2*n_aoa); % rad^2
cov_r = (err_tdoa_s*c)^2*eye(n_tdoa); % m^2
cov_rr = (err_fdoa_Hz*c/f_source_Hz)^2*eye(n_fdoa); % m^2/s^2
cov_z = blkdiag(cov_psi, cov_r, cov_rr);

% Generate noise
noise_white = randn(2*n_aoa+n_tdoa+n_fdoa, 1); % one row per sensor msmt
noise_sensor = chol(cov_z,'lower')*noise_white;    % apply error cov
```

Next, we generate the sensor reference indices. In this example, we're assuming 2-D AOA measurements, so each sensor will be referenced twice for AOA (first for azimuth measurements and then for elevation measurements).

```
N = size(x_sensor_enu,2);
[ref_tfdoa, test_tfdoa] = utils.parseReferenceSensor(N, N);
ref_idx  = [1:2*N,      2*N + ref_tfdoa,  3*N + ref_tfdoa];
test_idx = [nan(1,2*N), 2*N + test_tfdoa, 3*N + test_tfdoa];
```

With these, we can resample the noise and generate measurements for AOA (az/el), TDOA, and FDOA.

```
z = hybrid.measurement(x_sensor_enu, x_sensor_enu, x_sensor_enu,
    v_sensor_enu - v_source_enu, x_source_enu, ref_tdoa, ref_fdoa);
noise_msmt = utils.resampleNoise(noise_sensor,test_idx, ref_idx);
zeta = z + noise_msmt;
```

Solving is straightforward, and is followed with the same approach seen before in countless examples in this text and the last. For demonstration purposes, we apply the generic iterative least squares solution, starting with an initial solution just 1 km to the east and 1 km to the north of the reference sensor; the result is plotted in Figure 4.10.

```
x_init_enu = [1; 1; 0]*1e3;

epsilon = 100; % stopping condition (m)
max_num_iterations = 100; % stopping condition
```

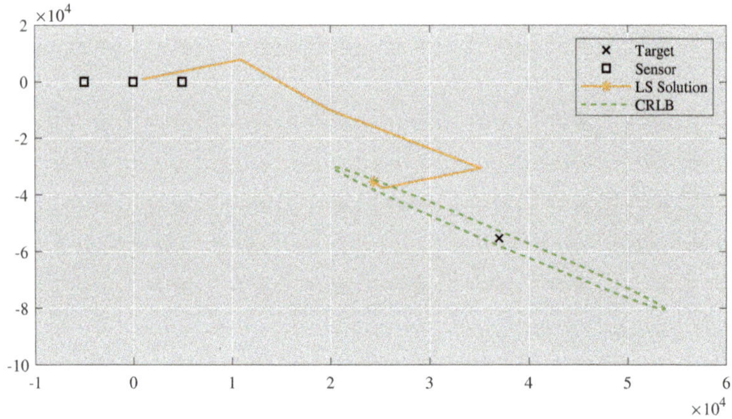

Figure 4.10 Plot of the laydown and solution for Example 4.1, in ENU coordinates (east-north shown, up not plotted). Dimensions are in meters.

```
force_full_calc = false;
plot_progress = false;

x_soln_enu = hybrid.lsSoln(x_sensor_enu, x_sensor_enu, x_sensor_enu,
    v_sensor_enu, zeta(:,idx), cov_x, x_init_enu, epsilon,
    max_num_iterations, force_full_calc, plot_progress, n_sensor, n_sensor)
    ;
```

To prepare our output, we convert the solution to LLA, and then from there to AER, to get an understanding of the miss distance from the true position.

```
[soln_lat, soln_lon, soln_alt] = utils.enu2lla(x_soln_enu(1), x_soln_enu(2),
     x_soln_enu(3), ref_lat, ref_lon, ref_alt);
[err_az, err_el, err_range] = utils.lla2aer(soln_lat, soln_lon, soln_alt,
    source_lat, source_lon, source_alt);
```

For the random noise we computed, the error resulted in 19.099 km of range between the solution and the true source position. For some statistical understanding, we repeated this process 100,000 times, removed outliers (those cases where the algorithm did not converge), and computed the RMSE in meters. As opposed to above, where we converted the output to LLA and then computed the miss distance, in this loop we show that the miss distance can be directly computed in ENU. The source position was previously converted to those units, thus preventing the need to convert the outputs to LLA (and then to AER) in order to compute the miss distance.

```
error = zeros(num_mc, max_num_iterations);

for idx=1:num_mc
```

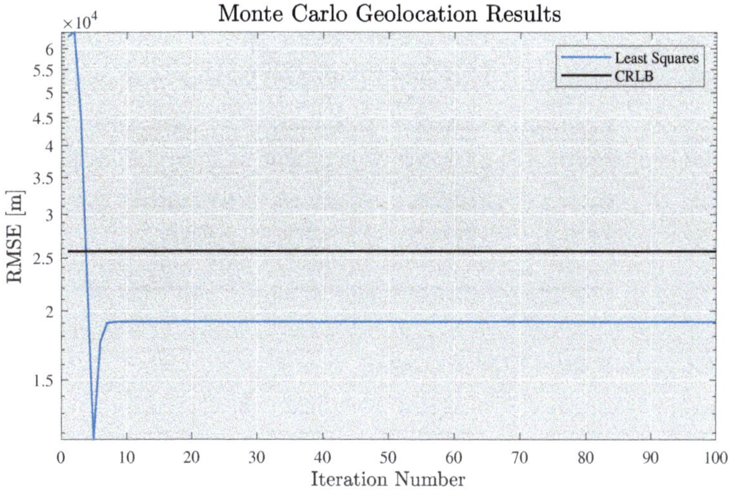

Figure 4.11 Comparison of the achieved RMSE using an interative LS solver for Example 4.1 with 10,000 Monte Carlo trials, to the RMSE predicted by the CRLB, both plotted in meters. Observe that only a few iterations are required in this geometry for convergence.

```
    [~, x_ls_iters] = hybrid.lsSoln(x_sensor_enu, x_sensor_enu, ...
        x_sensor_enu, v_sensor_enu, zeta(:,idx), cov_x, x_init_enu, ...
        epsilon, max_num_iterations, force_full_calc, plot_progress, ...
        ref_tdoa, ref_fdoa);

    error(idx,:) = sqrt(sum(abs(x_ls_iters-x_source_enu).^2,1));
end

% Remove outliers and compute RMSE
error = rmoutliers(error);
rmse = sqrt(sum(error.^2,1)/num_mc);
```

This result is then compared to the CRLB and plotted in Figure 4.11. In this case, we observe that the CRLB predicted RMSE of 25 km is achieved (on average) after only a few iterations.

```
crlb = hybrid.computeCRLB(x_sensor_enu, x_sensor_enu, x_sensor_enu, ...
                          v_sensor_enu, x_source_enu, cov_x);
rmse_crlb =sqrt(trace(crlb));
```

□

Table 4.2
Satellite Coordinates for Example 4.2

Satellite	Latitude	Longitude	Altitude (km)
Sat 1	26°52′ N	68°47′ W	500 km
Sat 2	27°52′ N	72°36′ W	500 km
Sat 3	23°19′ N	69°47′ W	500 km
Sat 4	24°19′ N	74°36′ W	500 km

Example 4.2 Satellite TDOA Example

Consider four satellites, whose positions (at a given point in time) are presented in Table 4.2. Compute the CRLB for points within a 500 km × 500 km region of the ocean, centered at 26°N, 71°W.

Solution: First, we define the satellites directly in LLA coordinates, and then convert them to ENU, at the reference location (the center of the CRLB image is a convenient reference to use).

```
% Center of observation area
ref_lat = 26; % deg Lat (N)
ref_lon = -71; % deg Lon (E)
ref_alt = 0; % alt (m)

[e_sensor, n_sensor, u_sensor] = utils.lla2enu(x_sensor_lla(1,:), ...
                                               x_sensor_lla(2,:), ...
                                               x_sensor_lla(3,:),...
                                               ref_lat, ref_lon, ref_alt);
```

Next, we define the grid of positions. Since the center of our grid is the origin in ENU coordinates, we can simply define the 500 km × 500 km grid relative to that point.

```
% Build grid of positions within 500km of source position (ENU origin)
xx_offset = linspace(-500e3,500e3,1001);

xx_grid = xx_offset;
yy_grid = xx_offset;
zz_grid = 0;

[XX,YY] = ndgrid(xx_grid,yy_grid);
x_grid = [XX(:), YY(:), zz_grid*ones(numel(XX),1)]';
```

Next, we define the sensor errors and handle the sensor pair assignments, as was discussed in Chapter 3.

```
n_tdoa = size(x_sensor_lla,2);
n_aoa = n_tdoa;
n_fdoa = n_aoa;
ref_tdoa = n_tdoa;
ref_fdoa = n_fdoa;

[tdoa_ref_vec,tdoa_tst_vec] = utils.parseReferenceSensor(ref_tdoa,n_tdoa);
[fdoa_ref_vec,fdoa_tst_vec] = utils.parseReferenceSensor(ref_fdoa,n_fdoa);
ref_idx = [1:2*n_aoa,      2*n_aoa+tdoa_ref_vec,2*n_aoa+n_tdoa+fdoa_ref_vec];
tst_idx = [nan(1,2*n_aoa),2*n_aoa+tdoa_tst_vec,2*n_aoa+n_tdoa+fdoa_tst_vec];
```

Finally, we use the indices to build the measurement covariance matrix, and compute the CRLB

```
cov_psi = (err_aoa_deg*pi/180)^2*eye(2*n_aoa); % rad^2
cov_r = (err_tdoa_s*utils.constants.c)^2*eye(n_tdoa); % m^2
cov_rr = (err_fdoa_Hz*utils.constants.c/f_source_Hz)^2*eye(n_fdoa);%(m/s)^2
cov_x = blkdiag(cov_psi, cov_r, cov_rr);

crlb = hybrid.computeCRLB(x_sensor_enu, x_sensor_enu, x_sensor_enu, ...
                          v_sensor_enu - v_source_enu, ...
                          x_grid, cov_x, ref_tdoa, ref_fdoa);
```

At this point, all that's left is calculation of the RMSE from the CRLB (shown below), and plotting (not shown). We could compute the RMSE with a for loop to iterate over the different source positions in the 500 $\times$ 500 km grid, but the arrayfun provides a convenient method to do so in a single line.

```
rmse_crlb = reshape(arrayfun(@(i) sqrt(trace(crlb(:,:,i))), 1:size(crlb,3)),
     size(XX));
```

We plot the result in Figure 4.12, and see the four circles indicating the ground track position of the four satellites. The color axis extends from 5 km to 10 km, showing us that pretty much the entire region enclosed by the satellites has 5–6 km of expected geolocation error, while it quickly increases to 10 km or greater error for points outside of them. □

The video linked in Figure 4.13 walks briefly through Example 4.2.

4.4 PROBLEM SET

4.1 Convert the position 34.0075°N, 118.500°W, at an altitude of 600 km to ECEF.

4.2 Convert the ECEF coordinates (-2519.333e3, -4659.311e3, 3541.351e3) to Lat/Lon/Alt.

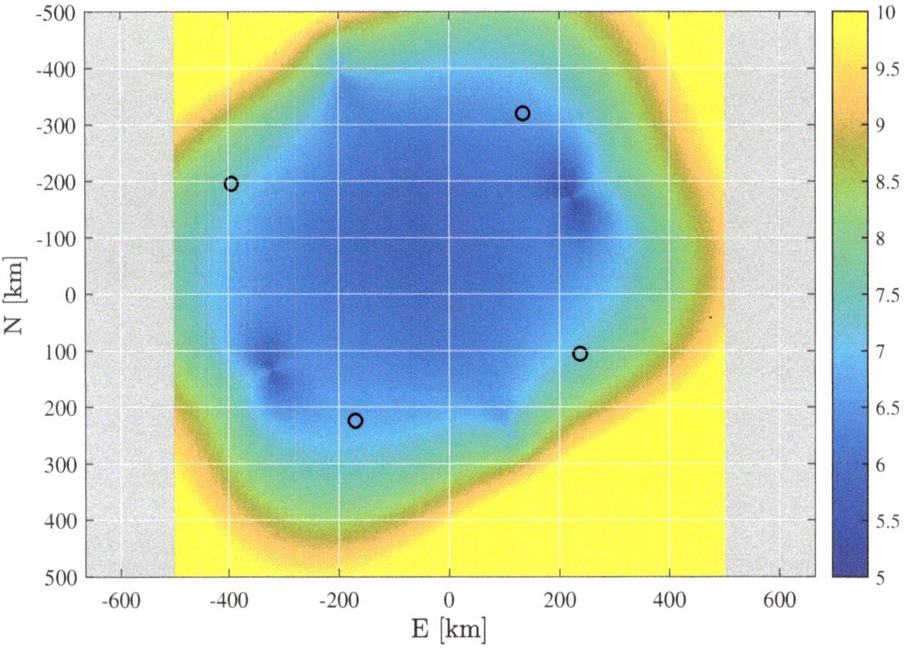

Figure 4.12 RMSE [km] computed from the CRLB for the 500 km × 500 km region defined in Example 4.2, in ENU coordinates (east-north shown, up not plotted). Note that the satellites (x) are 500 km above the plane where the CRLB is calculated.

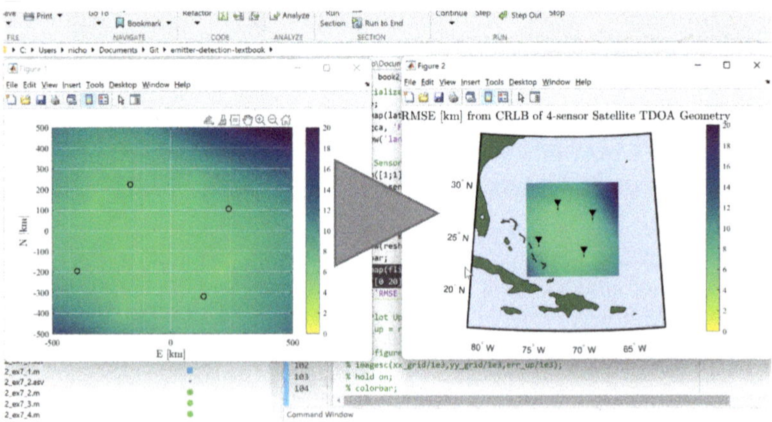

Figure 4.13 Video discussing Example 4.2.

4.3 Find the ECEF coordinates of the position 100 km NorthEast of 34.0075°N, 118.500°W, 30 m altitude. Convert to Lat/Lon/Alt.

4.4 Find the Lat/Lon coordinates of the position 213 km from 34.0075°N, 118.500°W, 30 m altitude, at an azimuth of 37°East of North, and an elevation angle of 10°.

4.5 Find the azimuth, elevation, and range of the point 33.944°N, 118.405°W, 100 m altitude, as observed from the point 34.0075°N, 118.500°W, 30 m altitude.

4.6 Find the East, North, Up coordinates of the point 33.944°N, 118.405°W, 100 m altitude, as observed from the point 34.0075°N, 118.500°W, 30 m altitude.

4.7 Repeat Example 4.2, but generate and plot the CRLB using Lat/Lon coordinates, rather than east-north. (Hint: if you generate a regular grid in East-North and then use imagesc to plot Lat/Lon, MATLAB® will assume the spacing is uniform. For large enough distances, this will distort the results. Generate the regular grid in Lat/Lon instead of east/north coordinates, convert to ENU to generate the CRLB, and then plot against the Lat/Lon vectors.)

References

[1] H. J. D'Souza and R. A. Bix, "Analytic geometry," *Encyclopedia Britannica*, 2016. [Online]. Available: https://www.britannica.com/science/analytic-geometry.

[2] DLR Earth Observation Center, *Imagery [basemap], web mapping service*, Oct. 2020. [Online]. Available: https://geoservice.dlr.de/web/services.

[3] J. R. Clynch, "Geodetic Coordinate Conversions," Naval Postgraduate School, Tech. Rep., 2002. [Online]. Available: https://www.oc.nps.edu/oc2902w/coord/coordcvt.pdf.

[4] D. A. Vallado and W. D. McClain, *Fundamentals of Astrodynamics and Applications*, 3rd. Microcosm Press, 2007.

[5] H. Moritz, "Geodetic Reference System 1980," *Bulletin Geodesique*, vol. 54, pp. 395–405, 1980. [Online]. Available: https://doi.org/10.1007/BF02521480.

[6] A. W. Doerry, "Earth Curvature and Atmospheric Refraction Effects on Radar Signal Propagation," Sandia National Laboratory, Sandia Report SAND2012-10690, 2013.

[7] "Department of Defense World Geodetic System 1984," National Imagery and Mapping Agency, Tech. Rep. TR 8350.2, Third Edition, Amendment 1, 2000.

[8] T. Burch, "Data collection of wgs84 info - or is it?" *GPS World*, 2016. [Online]. Available: https://www.gpsworld.com/data-collection-of-wgs-84-information-or-is-it/.

[9] "Recent update to wgs 84 reference frame and nga transition to igs antex," National Geospatial Agency, Office of Geomatics/GNSS Division, Tech. Rep., 2021. [Online]. Available: https://earth-info.nga.mil/php/download.php?file=(U)WGS%2084(G2139).pdf.

[10] J. R. Clynch, "Earth Models and Maps," Naval Postgraduate School, Tech. Rep., 2002. [Online]. Available: https://www.oc.nps.edu/oc2902w/general/mapmodel.pdf.

[11] United States Geological Service, *EarthExplorer*. [Online]. Available: https://earthexplorer.usgs.gov/.

[12] United States General Services Administration, *Data.gov Data Catalog*. [Online]. Available: https://catalog.data.gov/dataset.

[13] NASA Jet Propulsion Laboratory, *U.S. Releases Enhanced Shuttle Land Elevation Data*, 2014. [Online]. Available: https://www.jpl.nasa.gov/news/us-releases-enhanced-shuttle-land-elevation-data.

[14] United States National Oceanographic and Atmospheric Administration, *Latitude/Longitude Distance Calculator*. [Online]. Available: https://www.nhc.noaa.gov/gccalc.shtml.

[15] L. Herman, "The history, definition and pecularities of the earth centered inertial (eci) coordinate frame and the scales that measure time," in *1995 IEEE Aerospace Applications Conference. Proceedings*, vol. 2, 1995, pp. 233–263. DOI: 10.1109/AERO.1995.468933.

[16] G. L. Hall, "Compression of a radar track of a near Earth satellite into an Earth centered inertial state vector using least squares differential correction," Air Force Institute of Technology, Tech. Rep., 1994.

[17] Navigation and A. I. Facility, *An Overview of Reference Frames and Coordinate Systems in the SPICE Context*, 2020. [Online]. Available: https://naif.jpl.nasa.gov/pub/naif/toolkit_docs/Tutorials/pdf/individual_docs/17_frames_and_coordinate_systems.pdf.

[18] J. Zhu, "Conversion of earth-centered earth-fixed coordinates to geodetic coordinates," *IEEE Transactions on Aerospace and Electronic Systems*, vol. 30, no. 3, pp. 957–961, 1994. DOI: 10.1109/7.303772.

[19] D. K. Olson, "Calculation of geodetic coordinates from earth-centered, earth-fixed coordinates," *Journal of Guidance, Control, and Dynamics*, vol. 11, no. 2, pp. 188–190, 1988. DOI: 10.2514/3.56461. eprint: https://doi.org/10.2514/3.56461. [Online]. Available: https://doi.org/10.2514/3.56461.

[20] L. O. Lupash, "A new algorithm for the computation of the geodetic coordinates as a function of earth-centered earth-fixed coordinates," *Journal of Guidance, Control, and Dynamics*, vol. 8, no. 6, pp. 787–789, 1985.

[21] A. Nautiyal, "Algorithm to generate geodetic coordinates from earth-fixed coordinates," *Journal of Guidance, Control, and Dynamics*, vol. 11, no. 3, pp. 281–283, 1988. DOI: 10.2514/3.20306. eprint: https://doi.org/10.2514/3.20306. [Online]. Available: https://doi.org/10.2514/3.20306.

[22] Y. Wu, P. Wang, and X. Hu, "Algorithm of earth-centered earth-fixed coordinates to geodetic coordinates," *IEEE Transactions on Aerospace and Electronic Systems*, vol. 39, no. 4, pp. 1457–1461, 2003. DOI: 10.1109/TAES.2003.1261144.

[23] "Datum Transformations of GPS Positions," u-Blox, Application Note, 1999. [Online]. Available: https://microem.ru/files/2012/08/GPS.G1-X-00006.pdf.

Chapter 5

Geolocation with Target Constraints

In this chapter, we consider the problem of geolocating targets with constraints on the solution. This occurs when there is some foreknowledge of the target's position, such as a known altitude. We do not discuss in detail how this information might be obtained, as the methods are varied and outside the scope of this text. We will, however, consider two types of target constraints: hard and soft constraints.

Hard constraints take the form of fixed equality or inequality conditions and are expressed:

$$\begin{aligned} \mathbf{a}(\mathbf{x}) &= \mathbf{0} \quad \text{Equality constraints} \\ \mathbf{b}(\mathbf{x}) &\leq \mathbf{0} \quad \text{Inequality constraints} \end{aligned} \tag{5.1}$$

It is straightforward to implement a constrained solver that considers this information, if the constraints written above are convex [1].[1] Predicting estimation performance is slightly more difficult, since the Cramér-Rao lower bound (CRLB) does not consider the effect of constraints. For equality constrained conditions, we will use the *constrained* Cramér-Rao lower bound (CCRLB) to predict performance [2], [3], while we will show that inequality constraints have little effect on performance (unless the target is close to the inequality bound), and so we will estimate performance using the CRLB, even though it does not consider the effects of those constraints.

Soft constraints present statistical information about where the target is or is not likely to be located. Soft constraints can come from external measurements, such as a separate system's estimated target position, or prior estimates of a target's position and velocity (e.g., from a tracker). Soft constraints are represented as a prior probability distribution function $f_{\mathbf{X}}(\mathbf{x})$. In this manner, the posterior likelihood function for any set of measurements ζ is given:

$$\ell(\zeta, \mathbf{x}) = \ell(\zeta|\mathbf{x}) f_{\mathbf{X}}(\mathbf{x}). \tag{5.2}$$

1. Note that for convex optimization methods to be used, both the constraints and the *cost function* (in this case, the likelihood function or a simplified form of it) must also be convex. This is generally not true with geolocation problems, but they can often be assumed to be locally convex in a region around the true location $\mathbf{x}$.

where $\ell(\zeta|\mathbf{x})$ is the likelihood function, which computes how likely it is that the measurement ζ would occur if the true target location were $\mathbf{x}$. The posterior likelihood function $\ell(\zeta, \mathbf{x})$ reflects the combination of the measurement likelihood function and the a priori probability distribution function $f_{\mathbf{X}}(\mathbf{x})$.

The first step in applying target constraints is to translate them from whatever domain they were originally derived, to the solution domain. In our case, all of the algorithms we have developed, and will present in this text, operate on Cartesian coordinates, while constraints are often in geodetic coordinates. The most significant challenge is that while a constraint may be linear or convex in its original reference frame, once it is translated to a separate coordinate system, it may be nonlinear. For this reason, instead of a single generic derivation for all hard geographic constraints, we will specifically discuss two types of hard constraints: (1) fixed altitude (Section 5.1), and (2) bounded altitude (Section 5.2). Finally, we'll discuss the use of a priori estimates and information statistically in Section 5.3.

5.1 GEOLOCATION WITH KNOWN TARGET ALTITUDE

The first class of constraint is that of a known altitude. This case arises when an emitter is known to be on a ground or maritime vessel (and thus, at zero altitude), perhaps via recognition of the signal's content (such as an AIS broadcast).[2] It may also arise with a satellite emitter whose orbital period or velocity are known, both of which are tied directly to the orbital altitude.

Ho and Chan derive the problem of geolocation when the target altitude is known using TDOA and FDOA measurements in [6]. They show that, if the Earth is assumed to be spherical (and, therefore, the constraint has a straightforward representation in ECEF coordinates), then the problem can be solved directly. They also show that, due to the Earth's eccentricity, representation of a known altitude is more complex, and the solution requires an iterative approach. This built on earlier work by Niezgoda and Ho that solved for targets known to be on the Earth's surface, with the surface constraint expressed as a quadratic equation in ECEF coordinates, and solved the constrained optimization problem via Lagrange multipliers [7].

5.1.1 Optimization Problem Formulation

Geolocation can be represented as a constrained optimization problem where some constraint (or set of constraints) $\mathbf{a}(\mathbf{x}) = \mathbf{0}$ must be satisfied. We apply these constraints

2. AIS (the *automated identification system*) is a VHF-based transponder system for marine traffic that, while not intended when it was first deployed, is detectable from space [4], [5].

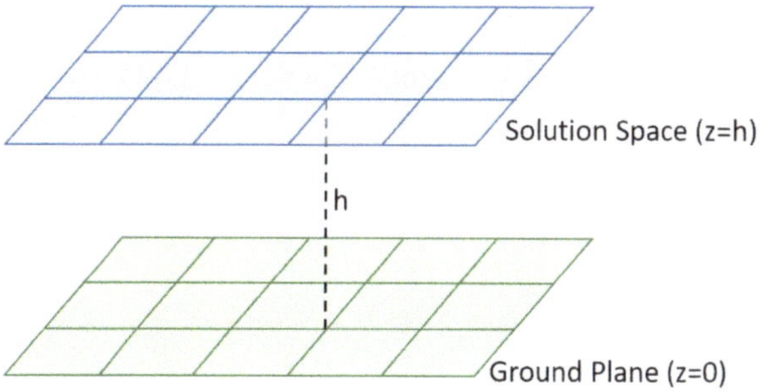

Figure 5.1 Illustration of a fixed altitude constraint for a flat-Earth model.

to the maximum likelihood formulation from (2.24).

$$\widehat{\mathbf{x}} = \arg\left[\max_{\mathbf{x}} \ell(\zeta|\mathbf{x})\right] \quad \text{subject to } \mathbf{a}(\mathbf{x}) = \mathbf{0} \tag{5.3}$$

What remains is to define the constraints $\mathbf{a}(\mathbf{x})$. In the simplest case, that of a flat Earth, a known altitude constraint $z = h$ is expressed directly with

$$a(\mathbf{x}) = \mathbf{x}^T [0,\ 0,\ 1] - h \tag{5.4}$$

This is not typically considered in the literature, given the inherent errors in a flat Earth assumption. Nevertheless, we include it here as a constraint valid for local problems expressed in ENU coordinates, for which the Earth's curvature is not taken into effect, and the fixed altitude constraint is equivalent to setting the up- (or z-) coordinate of the target location to $z = h$, as shown in Figure 5.1.

The next case, a spherical Earth, is shown in Figure 5.2, and results in a quadratic constraint based on the target's distance from the origin[6]:

$$a(\mathbf{x}) = \mathbf{x}^T\mathbf{x} - (R_e + h)^2 \tag{5.5}$$

where R_e is the average radius of the Earth, approximately 6,371 km (see Section 4.2.1.2).

If, however, the accuracy of a spherical Earth model is not sufficient, an ellipsoidal Earth model is required, as shown in Figure 5.3, and the fixed altitude constraint is much more complex [6]. Recall, from (4.12) that the altitude can be computed from ECEF coordinates via:

$$h = \frac{p}{\cos(\lambda)} - R_{\text{eff}}(\phi) \tag{5.6}$$

where $p = \sqrt{x^2 + y^2}$ is the distance from the center of the Earth to the point (x, y, z) in the equatorial plane defined in (4.8), ϕ is the geodetic latitude from (4.10), λ is the

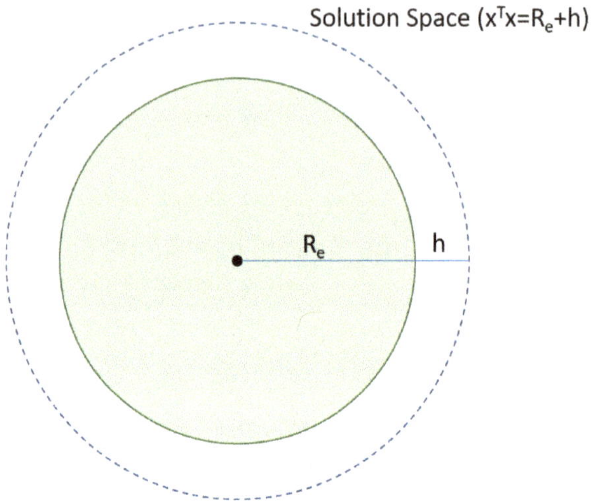

Figure 5.2 Illustration of a fixed altitude constraint for a spherical Earth model.

longitude from (4.11), and $R_{\text{eff}}(\phi)$ is the effective radius of the Earth at the latitude ϕ from (4.13). From this, we build out constraint

$$a(\mathbf{x}) = \frac{p}{\cos(\lambda)} - R_{\text{eff}}(\phi) - h \tag{5.7}$$

Figure 5.3 illustrates the difference between the geodetic distance calculation (the straight line through h and $R_e(\phi)$) and the geocentric distance calculation (the straight line to the origin).

These constraints can be generated with a call to `fixedAlt` in the package `utils.constraints`, which returns a pair of function handles, one for the desired constraint (defined above), and another for the gradient with respect to the target position $\mathbf{x}$ (defined in Section 5.1.3).

```
h = 10e3; % known target altitude
type = 'ellipse'; % {'flat', 'sphere', or 'ellipse'} are valid settings
[a, a_grad] = utils.constraints.fixedAlt(h, type);
```

5.1.2 Solvers

There are many ways to solve the optimization problem in (5.3). An analytic solution has not been derived, but many approximations and iterative solutions exist. We will first discuss how the solutions already presented in [8] can be adapted, then we will present a few select solutions that were derived specifically for the known altitude problem.

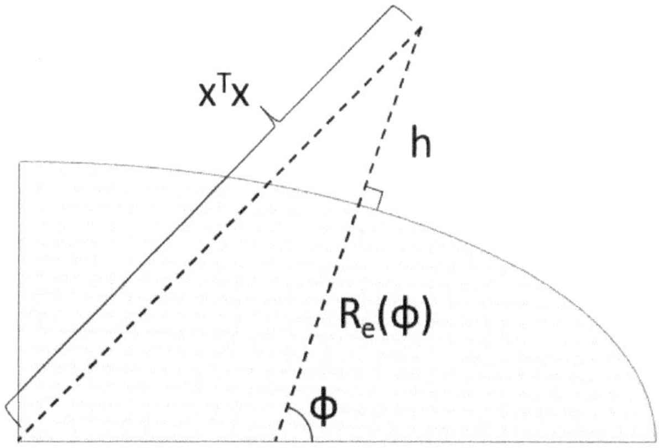

Figure 5.3 Illustration of a fixed altitude constraint for an ellipsoidal Earth model. Note that ϕ is the geodetic latitude, and $R_e(\phi)$ is the distance from the surface to the equatorial plane, at that latitude. The height constraint is also defined perpendicular to the Earth's surface. Note that the eccentricity of the Earth has been severely exaggerated in this graphic.

Solutions for this class of optimization problem (when the cost function and constraints are both convex) are discussed in Chapter 10 of [1].[3.] General solutions to this class of problem include a modification to Newton's approximation (used in the prior text as the basis for the convex solver) that considers the constraints, as well as several mathematical approaches to solving for an analytic solution.

5.1.2.1 Generic Solvers

In [8], we introduced three generic solvers based on the likelihood function $\ell(\mathbf{x}|\mathbf{z})$: (1) a brute-force solver that estimates the maximum likelihood solution, (2) an iterative least square solver, and (3) a gradient-descent based convex optimization solver. These methods can be directly leveraged to solve the constrained problem with three different approaches.

To modify the brute force ML solution, the solution space is directly constrained.

$$\ell_c(\zeta|\mathbf{x}) = \begin{cases} \ell(\zeta|\mathbf{x}) & \mathbf{a}(\mathbf{x}) = \mathbf{0} \;\&\; \mathbf{b}(\mathbf{x}) \leq \mathbf{0} \\ 0 & \text{else} \end{cases} \tag{5.8}$$

This constraint is achieved by passing the likelihood and constraints to the `constrainLikelihood` under `utils.constraints`, which returns a function handle

3. As was argued in [8], we will operate under a locally convex assumption, allowing us to leverage iterative solutions that work for convex problems even though the cost functions are not strictly convex; the principal risk of this approach is that the solver may converge to a false solution, particularly if the initial guess is too distant from the true solution.

for $\ell_c(\zeta|\mathbf{x})$. This approach is implemented in the function `mlSolnConstrained`, in each of the `aoa`, `tdoa`, `fdoa`, and `hybrid` packages. An example is shown below for FDOA, which demonstrates how to specify the constraints `a` and `b`, as well as the tolerance `tol` used for equality constraints.[4]

```
[x_est, A, grid] = fdoa.mlSolnConstrained(x_fdoa, v_fdoa, rho_dot, C,...
                        x_ctr,search_size,epsilon,a,b,tol);
```

For the iterative approaches, it is slightly more involved, but no less complex. We propose an iterative constrained algorithm, wherein we solve the unconstrained optimization problem iteratively, and then apply the constraint at the end of each iteration. To do so, we compute the scale term γ between the desired geocentric height of the target and the actual geocentric height at each iteration, then apply this scale term to the estimated position.[5]

$$\gamma^2 = R_d^2(\phi, h) / \mathbf{x}^{(i)T}\mathbf{P}\mathbf{x}^{(i)} \tag{5.9}$$

$$\widetilde{\mathbf{x}}^{(i)} = \gamma\mathbf{x}^{(i)} \tag{5.10}$$

where $\mathbf{x}^{(i)}$ is the ith iterative target location estimate (e.g., from gradient descent discussed in Section 2.2.3 or iterative least squares discussed in Section 2.2.4), and $\widetilde{\mathbf{x}}^{(i)}$ is the projection of that estimate onto the fixed altitude constraint. To facilitate this iterative solution/projection, the fixed constraint functions `fixedAlt` and `fixedCartesian`[6] return two outputs, the distance ϵ from the constraint, as well as the projection $\widetilde{\mathbf{x}}$ of the input position onto the constraint.

Within each class of problem (`aoa`, `tdoa`, `fdoa`, and `hybrid`), a pair of constrained solvers are provided, which implement (5.9-5.10), available as the functions `gdSolnFixed` and `lsSolnFixed`. They can be called with, for example:

```
[x, x_full] = hybrid.lsSolnFixed(x_aoa, x_tdoa, x_fdoa, v_fdoa, z, C, x_init
    , a, tol);
```

The new inputs are placed after the initial target position estimate `x_init` (this placement is the same for both the GD and LS solvers). `a` is the function handle for equality constraint,[7] and `tol` is an optional tolerance to use when determining if an equality

4. Instead of testing for strict equality with zero, the equality constraints are constrained to have a magnitude less than or equal to `tol`. This is more numerically stable when testing floating point numbers, which are almost never exactly equal to zero.
5. γ^2 in (5.9) is defined for elliptical problems; in the spherical case $R_{\text{eff}}(\phi)$ is replaced with R_e, and $\mathbf{P}$ is removed.
6. The `fixedCartesian` function is similar to the `fixedAlt` constraint builder that we previously defined, except that it can be used for generic constraints on a Cartesian plane, such as a fixed x, y, or z value, or a line defined by a point $\mathbf{x}_0$ and a vector $\mathbf{u}$. Both are found in the package `utils.constraints`.
7. If there is more than one constraint, they are passed in as a cell array of function handles.

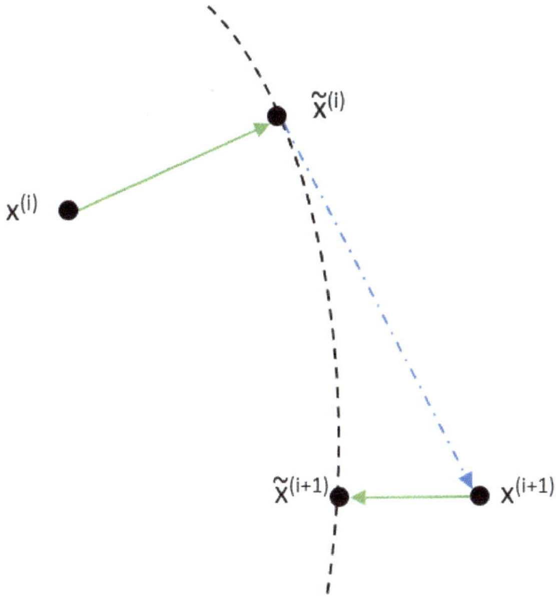

Figure 5.4 Illustration of the iterative constrained optimization approach from (5.9–5.10), with the blue dash-dot line representing the unconstrained iterative solution, and the green solid line representing projection onto the constrained solution space, which is represented by the black dashed line.

constraint is met or not. After the `tol` input is where the normal solver flags (such as the desired resolution `epsilon` or the flag `force_full_calc`) occur.

This iterative process is illustrated in Figure 5.4, for a simple 2-D case showing two iterations. The solid line is the projection of each estimate onto the constraint space (represented by the dashed line), while the dash-dot line is the next iterative (unconstrained) update.

These implementations are meant to represent the standard approach to constraining the optimization problem, and do not reflect any improvements (either for convergence speed or stability) that can be achieved with knowledge of the type of constraint in use. Such improvements are left to the readers as an exercise, and to researchers furthering the field of study, in which case these algorithms can be used as a benchmark.

Example 5.1 2-D Bearing-Only Geolocation with Known Range

As a simplified example, let us first consider a 2-D problem with two AOA sensors at (−2, 0) and (2, 0). The first measures an angle of arrival on the target of 80°, and the

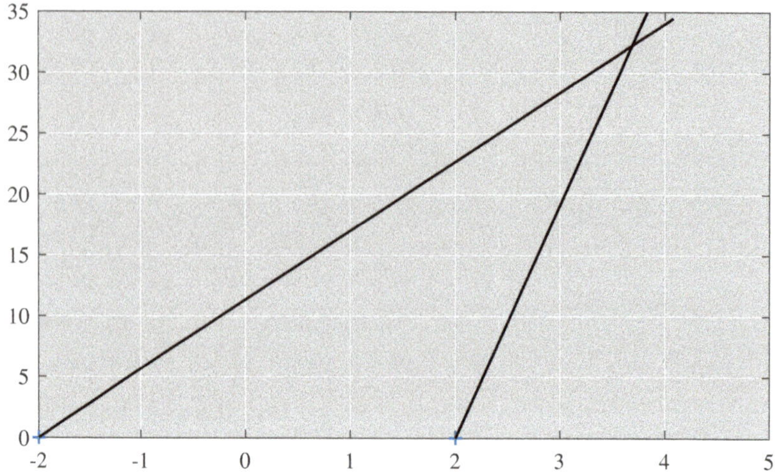

Figure 5.5 Layout for Example 5.1 with two bearing-only sensors.

latter 87°.[8] Plot the scenario, and use the gradient descent solver to geolocate the target. Assume that the two sensors have equal noise performance ($\mathbf{C} = \sigma^2\mathbf{I}$).

Then, assume that the down-range position y is known to be 25. Generate a constraint function, and use it to solve for x with the fixed constraint gradient descent solver. Plot both solutions, along with the sensors and lines of bearing.

Solution: The first portion is very straightforward; we define the sensors and their measurements, and plot the scenario in Figure 5.5. We can see from Figure 5.5 that the solution should be near the point (3.7, 32), based on a visual approximation. Although the covariance was unspecified, we don't need absolute knowledge of $\mathbf{C}$ in order to execute the solver, we only need to know that all the measurements are equally noisy (and thus, that they should all be equally weighted in the solution). Thus, we define $\mathbf{C} = \mathbf{I}$ for simplicity. We also must provide the GD solver with an initial position estimate, so we choose a point between the two sensors and slightly ahead as our seed.

```
x_aoa = [-2, 2; 0, 0];
psi = [80; 87]*pi/180;
C = eye(n_aoa); % Make it 1 radian std. dev.
x_init = [0;1]; % initial guess
```

The solution is quickly computed with a call to `triang.gdSoln`.

```
[x_ls, x_ls_full] = triang.gdSoln(x_aoa, psi, C, x_init);
```

8. Recall that the `triang.measurement` function defines bearing as radians CCW from +x.

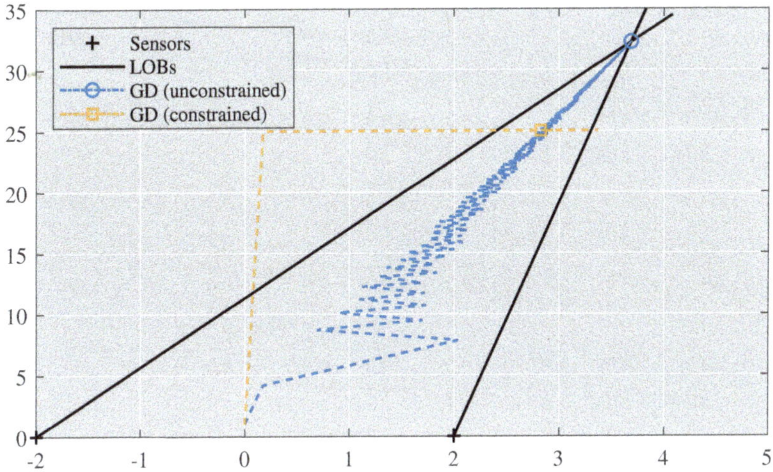

Figure 5.6 Layout and solution for Example 5.1 with two bearing-only sensors, an unconstrained gradient descent solution, and a constrained ($y = 25$) gradient descent solution.

To construct the constrained solution, we first define our constraint, $y = 25$, which we can see from Figure 5.5 is not where the intersection is. So, we expect that our constrained solution will be different than the unconstrained. To define our constraint, we use the function `fixedCartesian` of the `utils.constraints` package. Once the constraint is constructed, we supply it to the solver `gdSolnFixed` in the same way that we would with `gdSoln`.

```
y_soln = 25;
[a, ~] = utils.constraints.fixedCartesian('y',y_soln);
[x_gd_const,x_gd_full_const] = triang.gdSolnFixed(x_aoa, psi, C, ...
                                        x_init, a);
```

The results for both the constrained and unconstrained are plotted in Figure 5.6 and printed here:

```
Unconstrained Solution: (3.69, 32.28)
Constrained Solution:   (2.83, 25.00)
```

The unconstrained solution follows a traditional gradient descent approach (whose oscillating behavior we've shown before) and converges to the intersection of the two LOBs, as we should expect. The constrained solution, however, quickly snaps to $y = 25$ on the first iteration, and then traverses along that line toward the most likely solution, which appears roughly between the two LOBs. This is the most likely solution that satisfies our fixed constraint. □

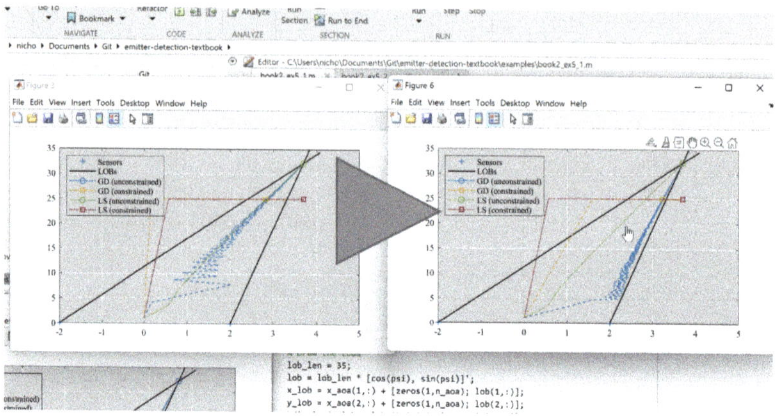

Figure 5.7 Video discussing Example 5.1, the ability of the constrained LS to solve the problem, and the impact of nonuniform sensor performance.

The video linked in Figure 5.7 discusses Example 5.1 with some modifications. First, we discuss the ability of the constrained LS solver to find the solution. Then, we modify the assumed sensor performance and discuss how both the constrained GD and constrained LS solvers behave.

Example 5.2 Airborne TDOA Sensors Geolocating a Ground-Based Target

Consider four TDOA sensors, at 1 km altitude, arranged in an east-west line, with 10 km spacing between each sensor. There is an emitter 50 km north and 10 km west of the western-most sensor, at an altitude of 100 m. Assume all sensors have a timing accuracy of 100 ns.

Generate a case with random sensor noise, and solve it using the iterative gradient descent solver both with and without knowledge of the target's altitude (100 m). In both cases, initialize the solution with a guess of 0 km E, 10 km N, and coaltitude with the sensors (1 km altitude). Plot the sensors, target, and both sets of solutions. Comment on the trajectory and accuracy.

Solution: To begin, we define the sensors and target, as well as the TDOA measurement error and measurements, using approaches that have been covered countless times in this text.

```
alt = 1e3;
x_tdoa = [-15e3, -5e3, 5e3, 15e3;
          0, 0, 0, 0;
          alt1, alt1, alt1, alt1];
[~,n_tdoa] = size(x_tdoa);
```

```
% Define target position and initial estimate
tgt_alt = 100; % known target altitude
x_tgt = [-10e3; 40e3; tgt_alt];
x_init = [0;10e3;alt1];

% Sensor Accuracy
time_err = 1e-7;
Croa = (utils.constants.c*time_err)^2 * eye(n_tdoa);
L = chol(Croa,'lower');

% Measurement and Noise
z = tdoa.measurement(x_tdoa, x_tgt, []);
noise = L*randn(n_tdoa,1);               % generate sensor noise
noise_z = utils.resampleNoise(noise, []); % generate measurement noise
zeta = z + noise_z;
```

To solve for the target position both without the altitude constraint, we use `tdoa.gdSoln`, and with the altitude constraint, we use `tdoa.gdSolnFixed` with the constraint provided by `utils.constraints.fixedAlt`.

```
[x_gd, x_gd_full] = tdoa.gdSoln(x_tdoa, zeta, Croa, x_init);

[a, ~] = utils.constraints.fixedAlt(tgt_alt, 'flat');
[x_gd_alt, x_gd_alt_full] = tdoa.gdSolnFixed(x_tdoa, zeta, Croa, x_init, a);
```

The result for these two cases, and the convergence of each solver, is plotted in Figure 5.8. The two final solutions are:

```
Unconstrained Solution: -9.87 km E, 39.64 km N, 1.00 km U
Constrained Solution:   -9.87 km E, 39.63 km N, 0.10 km U
```

We can see here that the unconstrained solver did very well to estimate the east and north coordinates, likely due to the low measurement noise and good spacing of the sensors in the east-west dimension. The unconstrained solution, however, did not deviate at all from the initial estimate of 1 km for the altitude. This is primarily because the lack of any altitude variation in the sensors means that there is no information provided in the up dimension, and thus the gradient descent solver has no reason to adjust this estimate. This is observed in Figure 5.8, where we see that the unconstrained solver never deviates from this initial altitude. The constrained solver, however, immediately selects the correct altitude. One note we will make is that the down-range (north) position estimate is slightly more accurate at the true altitude of 100 m. In the east and north dimensions, we see that the two estimates are almost identical, deviating by approximately 10 m in the north dimension, and both within 400 m of the true target position. □

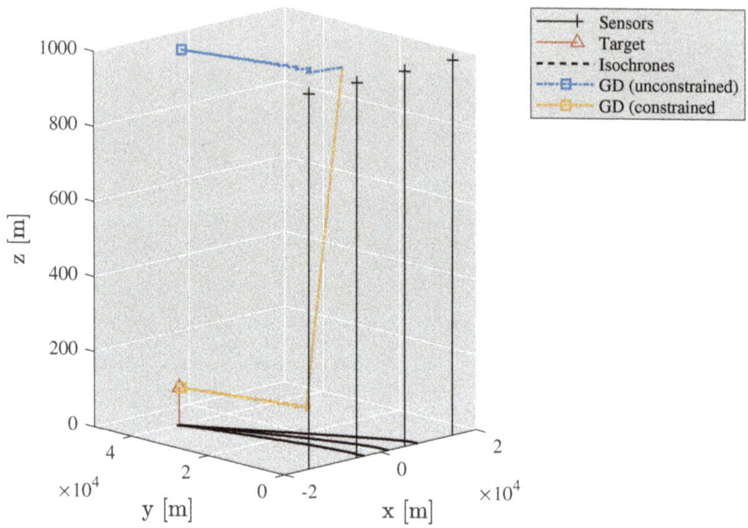

Figure 5.8 Laydown and solution for Example 5.2. Isochrones (at z=0) are drawn in black, for reference.

In the video linked in Figure 5.9, we explore how the use of elevation diversity can improve accuracy.

5.1.2.2 Other Solutions

We briefly introduce several other solutions, although they are not implemented in the associated software for this text. Bin et al. [9] use particle swarm optimization to solve the constrained localization problem using a combination of TDOA, FDOA, and AOA measurements. Cao et al. [10] use a novel approximate cost function for TDOA/FDOA estimation, similar to a constrained maximum likelihood estimation approach, which yields an iterative solver for known altitude targets (and also considers sensor location uncertainty).

Chan and Ho showed that their algebraic TDOA solution (described in Chapter 11 of [8], and implemented in `tdoa.chanHoSoln`) could be extended to account for a known altitude constraint [6]. Interested readers are encouraged to explore that paper for a discussion of several special cases, including an algebraic solution when exactly three sensors are available and an iterative solution using Lagrange multipliers for overdetermined cases, or to Prof. Ho's laboratory website for MATLAB® code implementing this and many other relevant geolocation algorithms.

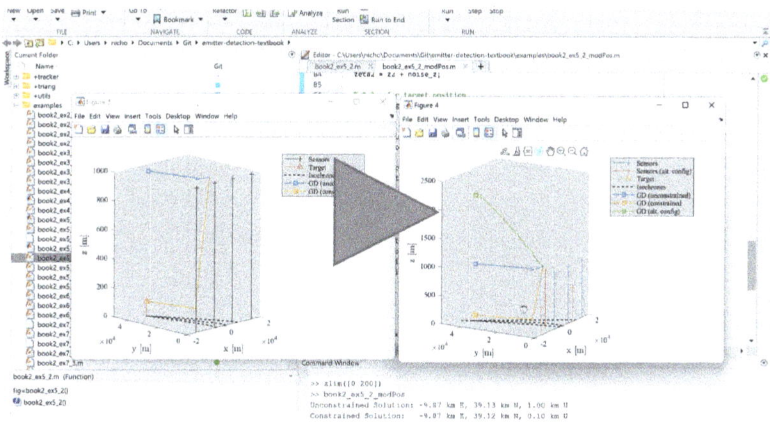

Figure 5.9 Video discussing Example 5.2 and the use of elevation diversity among the sensors to achieve elevation accuracy in the geolocation estimate.

Another general class of solutions, to the problem of an equality constrained optimization problem, is to encode the constraint directly onto the cost function via Lagrange multipliers [1]. In our case, this would take the form:[9]

$$\ell((\mathbf{x}|\boldsymbol{\zeta}, \mathbf{a}(\mathbf{x}), \lambda) = \ell(\boldsymbol{\zeta}|\mathbf{x}) - \lambda a^2(\mathbf{x}) \tag{5.11}$$

By including this second term (with $\lambda > 0$, and taking the square of the difference between the desired and actual altitude), we impose any deviation from the known altitude constraint as a reduction in the likelihood function. If the nonlinear constraint $a(\mathbf{x})$ is differentiable, then this modification lends itself to the numerical methods used to solve convex problems, as we've applied to other geolocation problems in this text and its predecessor. The principal challenge with this approach is the Lagrange parameter (λ). Most solutions rely on an iterative approach to compute the optimal value of λ, such as a Newton's method search, then use that optimized λ to solve the estimation problem with (5.11).

5.1.3 Performance Analysis

To analyze performance, we will look to the constrained CCRLB. This is an extension of the CRLB that accounts for the reduced uncertainty that comes with constraining the solution space. It was originally derived by Gorman and Hero [2], [3], but has been

9. Note that in the general convex optimization literature, the standard form is a minimization of some cost function, whereas in our formulation, we are maximizing a likelihood function. This is why we subtract the Lagrange term instead of adding it.

shown to be a simple modification of the traditional CRLB [11].

$$\mathbf{C}_\mathbf{x} \geq \mathbf{F}_\mathbf{z}^{-1}(\mathbf{x}) - \mathbf{F}_\mathbf{z}^{-1}(\mathbf{x})\dot{\mathbf{A}}\left(\dot{\mathbf{A}}^T\mathbf{F}_\mathbf{z}^{-1}(\mathbf{x})\dot{\mathbf{A}}\right)^{-1}\dot{\mathbf{A}}^T\mathbf{F}_\mathbf{z}^{-1}(\mathbf{x}) \tag{5.12}$$

where the matrix $\mathbf{F}_\mathbf{z}^{-1}(\mathbf{x})$ is the inverse of the Fisher information matrix (for the unconstrained problem), and $\dot{\mathbf{A}}$ is the gradient of the set of (possibly nonlinear) constraints ($\mathbf{a}(\mathbf{x})=\mathbf{0}$), with respect to each of the estimated variables

$$\dot{\mathbf{A}} = \nabla_\mathbf{x}\mathbf{a}^T(\mathbf{x}) = \left[\nabla_\mathbf{x}a_0(\mathbf{x}), \dots, \nabla_\mathbf{x}a_{N-1}(\mathbf{x})\right] \tag{5.13}$$

where $\mathbf{a}(\mathbf{x}) = [a_0(\mathbf{x}), \dots, a_{N-1}(\mathbf{x})]$ is the set of equality constraints. The first term of the CCRLB is simply the CRLB, and shows that the error performance for the constrained optimization case begins with the unconstrained. The second term of the CCRLB (after the subtraction) represents the reduction in uncertainty provided by the constraints. It is straightforward to show that this term is positive-semidefinite and, therefore, that the CCRLB is strictly not larger than the CRLB. In other words, applying an altitude constraint will (assuming it is correct) never increase the expected error of the solution.

$$\mathbf{C}_\mathbf{x} \geq \underbrace{\mathbf{F}_\mathbf{z}^{-1}(\mathbf{x})}_{\text{CRLB}} - \underbrace{\mathbf{F}_\mathbf{z}^{-1}(\mathbf{x})\dot{\mathbf{A}}\left(\dot{\mathbf{A}}^T\mathbf{F}_\mathbf{z}^{-1}(\mathbf{x})\dot{\mathbf{A}}\right)^{-1}\dot{\mathbf{A}}^T\mathbf{F}_\mathbf{z}^{-1}(\mathbf{x})}_{\text{Effect of constraints}} \tag{5.14}$$

5.1.3.1 Spherical Earth CCRLB

In the case of known altitude for a spherical Earth problem, the gradient of the constraint function simplifies to $\dot{\mathbf{A}} = \mathbf{x}$, and the CCRLB is given [6]:

$$\mathbf{C}_\mathbf{x} \geq \mathbf{F}_\mathbf{z}^{-1}(\mathbf{x}) - \frac{1}{\mathbf{x}^T\mathbf{F}_\mathbf{z}^{-1}(\mathbf{x})\mathbf{x}}\mathbf{F}_\mathbf{z}^{-1}(\mathbf{x})\mathbf{x}\mathbf{x}^T\mathbf{F}_\mathbf{z}^{-1}(\mathbf{x}) \tag{5.15}$$

5.1.3.2 Ellipsoidal Earth CCRLB

For an elliptical Earth, the gradient $\dot{\mathbf{A}}$ is much more involved. To begin, we take the partial derivative of the constraint from (5.7) with respect to x, y, and z:

$$\frac{\partial a(\mathbf{x})}{\partial x} = \frac{x^2 - y^2}{x\sqrt{x^2+y^2}\cos(\lambda)} - \frac{\partial R_{\text{eff}}(\phi)}{\partial x} \tag{5.16}$$

$$\frac{\partial a(\mathbf{x})}{\partial y} = \frac{x + y}{\sqrt{x^2+y^2}\cos(\lambda)} - \frac{\partial R_{\text{eff}}(\phi)}{\partial y} \tag{5.17}$$

$$\frac{\partial a(\mathbf{x})}{\partial z} = -\frac{\partial R_{\text{eff}}(\phi)}{\partial z} \tag{5.18}$$

where $\beta = 1 - e_1^2$ is in shorthand to save space. The partial derivatives of the effective radius with respect to x, y, and z are given:

$$\frac{\partial R_{\text{eff}}(\phi)}{\partial x} = \frac{ae_1^2 \sin(\phi)\cos(\phi)}{\left(1 - e_1^2 \sin^2(\phi)\right)^{3/2}} \frac{\partial \phi}{\partial x} \tag{5.19}$$

$$\frac{\partial R_{\text{eff}}(\phi)}{\partial y} = \frac{ae_1^2 \sin(\phi)\cos(\phi)}{\left(1 - e_1^2 \sin^2(\phi)\right)^{3/2}} \frac{\partial \phi}{\partial y} \tag{5.20}$$

$$\frac{\partial R_{\text{eff}}(\phi)}{\partial z} = \frac{ae_1^2 \sin(\phi)\cos(\phi)}{\left(1 - e_1^2 \sin^2(\phi)\right)^{3/2}} \frac{\partial \phi}{\partial z} \tag{5.21}$$

and the partial derivatives of the latitude (ϕ), with respect to x, y, and z are given:

$$\frac{\partial \phi}{\partial x} = \frac{-xz\beta}{\sqrt{x^2+y^2}\left(z^2+\beta^2\left(x^2+y^2\right)\right)} \tag{5.22}$$

$$\frac{\partial \phi}{\partial y} = \frac{-yz\beta}{\sqrt{x^2+y^2}\left(z^2+\beta^2\left(x^2+y^2\right)\right)} \tag{5.23}$$

$$\frac{\partial \phi}{\partial z} = \frac{\beta\sqrt{x^2+y^2}}{z^2+\left(1-e_1^2\right)\left(x^2+y^2\right)} \tag{5.24}$$

This can be computed numerically by using the provided `fixedAlt` constraint generator in the `utils.constraints` package.[10.]

```
[a, a_grad] = utils.constraints.fixedAlt(h, type);
```

and passing the second function handle (`a_grad`) to the new constrained CRLB function in the relevant solver space (`aoa`, `tdoa`, `fdoa`, or `hybrid`):

```
crlb = hybrid.constrainedCRLB({}, a_grad);
```

where `{}` is the list of parameters required by the corresponding unconstrained CRLB. The function handle `a_grad` accepts an `nDim x M` set of emitter positions and returns the gradient of the fixed altitude constraint at each position. The constrained CRLB then uses this return to compute (5.12).

Example 5.3 Constrained CRLB of a Hybrid Geolocation System

Consider a hybrid geolocation system, with a ground-based AOA sensor (0 m alt), and two TDOA receivers due east of the AOA sensor at 20 km and 25 km, also at 0 m altitude. The emitter is located at a ground range of 50 km from the AOA sensor, at a

10. Valid settings for the constraint type are `'flat'`, `'sphere'`, or `'ellipse'`.

bearing of 30° east from north, and an altitude of 10 km. The AOA sensor has a 5-degree angle error in both azimuth and elevation, and the TDOA sensors have a $1-\mu$s error.

Compute the CRLB at the true target position both with and without knowledge of the target altitude. Compare the results.

Plot the RMSE, computed from the CRLB for a grid of 101 × 101 points spanning 20 km E × 20 km N, centered on the true target location, both with and without knowledge of the target altitude.

Solution: First, as always, we define the sensor and target positions.

```
x_aoa = [0, 0, 0]'; % AOA sensor at reference LLA; origin in ENU
x_tdoa = [20e3, 25e3; 0, 0; 0, 0];
n_aoa = size(x_aoa,2); n_tdoa = size(x_tdoa,2);

bng_tgt = 30; % deg E of N
tgt_rng = 50e3;
tgt_alt = 10e3;
x_tgt = [tgt_rng * sind(bng_tgt);
         tgt_rng * cosd(bng_tgt);
         tgt_alt];
```

We define the angle and time of arrival accuracies, use them to construct covariance matrices for AOA and TDOA, and combine them to form the full sensor covariance matrix. Even though there is only one AOA sensor, we generate a 2 × 2 covariance matrix for AOA, because it generates both azimuth and elevation measurements (to turn that off, we can use 2-D positions, or pass a flag to `hybrid.computeCRLB` that specifies `do2DAoA=false`).

```
err_aoa = 3*pi/180; % 1 deg
err_toa = 1e-6;
C_aoa = err_aoa^2*eye(2);
C_roa = (utils.constants.c*err_toa)^2*eye(n_tdoa);
C_full = blkdiag(C_aoa, C_roa);
```

The actual usage of these parameters to compute the CRLB and constrained CRLB is very straightforward. The second output of the call to `fixedAlt` is a function handle that computes the gradient of the constraints, according to (5.13). Then, it's a simple matter of passing the same inputs to `computeCRLBfixed` along with `a_grad`, which then implements the CCRLB defined in Section 5.1.3.

```
C_raw = hybrid.computeCRLB(x_aoa, x_tdoa, [], [], x_tgt, C_full);

[~, a_grad] = utils.constraints.fixedAlt(tgt_alt, 'flat');
C_fix = hybrid.computeCRLBfixed(x_aoa,x_tdoa,[],[],x_tgt,C_full,a_grad);
```

The result is printed as a pair of 3 × 3 covariance matrices. The former (unconstrained) shows errors as much as $8 * 10^7$ m^2, while the principal error in the latter is only $5.7 * 10^7$ m^2. More importantly, the third dimension (altitude) has no noise in the constrained case, which verifies the expected performance, that constraining altitude means there will be no error in that dimension (assuming that the constraint is accurate). Interestingly, this also translates to reduced error in the other dimensions, not merely a projection of the covariance matrix onto the 2×2 subspace that remains unconstrained. This improvement is driven by the fact that now all of the information gathered by the sensors can be leveraged to solve fewer unknowns, leading to increased accuracy in the remaining estimates.

```
CRLB (unconstrained):
   1.0e+07 *

    1.7851    3.2650    0.7544
    3.2650    8.6964    1.8508
    0.7544    1.8508    1.1415

CRLB (constrained):
   1.0e+07 *

    1.2865    2.0418         0
    2.0418    5.6956         0
         0         0         0
```

Finally, we repeat this measurement over a grid of positions centered on the target, rather than just the central point. The code is very similar to before, once we define the grid of test points. The one difference is that, to support plotting, we must collapse each 3×3 covariance matrix to a single metric. We use the RMSE, computed as the square root of the trace of each covariance matrix. The result is plotted in Figure 5.10.

```
% Plot for x/y Grid
xy_vec = linspace(-10e3, 10e3, 201);
[X,Y] = meshgrid(xy_vec);
x_grid = [X(:), Y(:), zeros(numel(X),1)]' + x_tgt;

% Define target position and initial estimate
C_raw_grid = hybrid.computeCRLB(x_aoa, x_tdoa, [], [], x_grid, C_full);
C_fix_grid = hybrid.computeCRLBfixed(x_aoa, x_tdoa, [], [], x_grid, C_full,
    a_grad);

% Compute RMSE
rmse_raw = reshape(sqrt(C_raw_grid(1,1,:)+C_raw_grid(2,2,:)+C_raw_grid
    (3,3,:)),size(X));
rmse_fix = reshape(sqrt(C_fix_grid(1,1,:)+C_fix_grid(2,2,:)+C_fix_grid
    (3,3,:)),size(X));
```

□

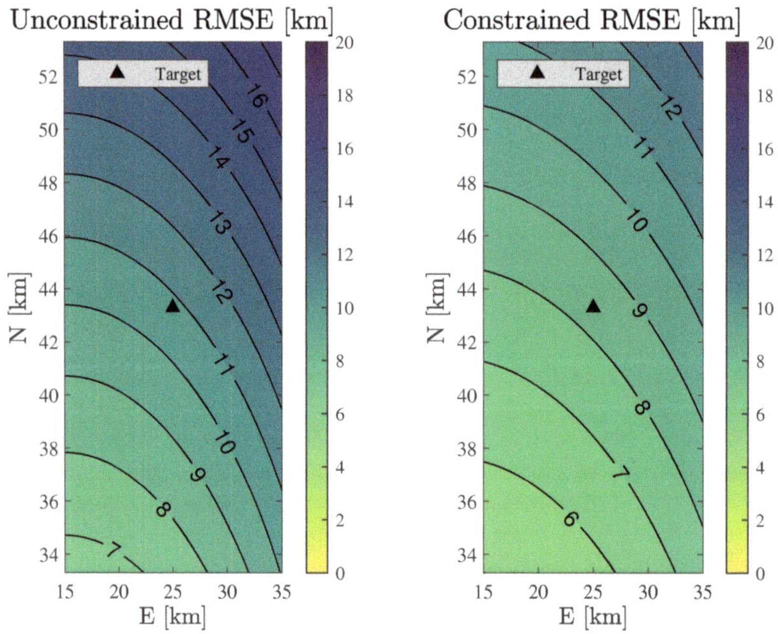

Figure 5.10 Geolocation performance, modeled with the RMSE computed from the CRLB with and without altitude constraint, for Example 5.3.

5.2 GEOLOCATION WITH KNOWN ALTITUDE BOUNDS

In some cases, it is feasible that a target's altitude may be bounded but not known explicitly. To repeat the satellite example, it is somewhat more likely that a satellite's orbital altitude may be known with some error, perhaps because a launch vehicle was observed but that some allowance must be made for maneuvers after separation of the satellite, or similarly that an allowance must be made following its most recent observation.[11] For the terrestrial case, we could consider that a maritime vessel may have its antenna high on a mast of unknown height, and so the altitude should not be strictly zero, but merely close to the surface. Airborne targets may similarly be known to occupy a specific band, whether because civilian governments restrict the allowable altitudes, or because intelligence data provides some knowledge of where a given aircraft can be expected to operate.[12] In any of these cases, consideration of the known altitude bounds may increase geolocation accuracy and will certainly not decrease it.

5.2.1 Optimization Problem Formulation

Similar to the known altitude case, we present the constrained optimization problem, this time with inequality constraints.

$$\begin{aligned}\widehat{\mathbf{x}} &= \arg\max_{\mathbf{x}} \ell(\mathbf{x}|\mathbf{z}) \\ &\text{subject to } \mathbf{b}(\mathbf{x}) \leq \mathbf{0}\end{aligned} \tag{5.25}$$

What remains is to define the inequality constraints $\mathbf{b}(\mathbf{x})$. In the simplest case, that of a flat Earth, a known altitude constraint $z \geq h_{\text{min}}$ and $z \leq h_{\text{max}}$ is shown in Figure 5.11 and is expressed:

$$\mathbf{b}(\mathbf{x}) = \begin{bmatrix} \mathbf{x}^T [0,\ 0,\ 1] - h_{\text{max}} \\ h_{\text{min}} - \mathbf{x}^T [0,\ 0, 1] \end{bmatrix} \tag{5.26}$$

We can similarly express altitude bounds for a spherical Earth:

$$\mathbf{b}(\mathbf{x}) = \begin{bmatrix} \mathbf{x}^T\mathbf{x} - (R_e + h_{\text{max}})^2 \\ (R_e + h_{\text{min}})^2 - \mathbf{x}^T\mathbf{x} \end{bmatrix} \tag{5.27}$$

and for an ellipsoidal Earth:

$$\mathbf{b}(\mathbf{x}) = \begin{bmatrix} \mathbf{x}^T\mathbf{P}\mathbf{x} - (R_{\text{eff}}(\phi) + h_{\text{max}})^2 \\ (R_{\text{eff}}(\phi) + h_{\text{min}})^2 - \mathbf{x}^T\mathbf{P}\mathbf{x} \end{bmatrix} \tag{5.28}$$

11. For a discussion on the challenge of tracking satellites, including the impact of measurement errors on predictions for later passes and atmospheric drag on low altitude orbits, see, for example, [12]–[15].
12. For a discussion of how difficult airspace deconfliction can be, see this report on Operation Iraqi Freedom [16]. Some articles on approaches to improving airspace deconfliction, which include the use of commanded altitude bands for various aircraft missions, can be found in [17], [18].

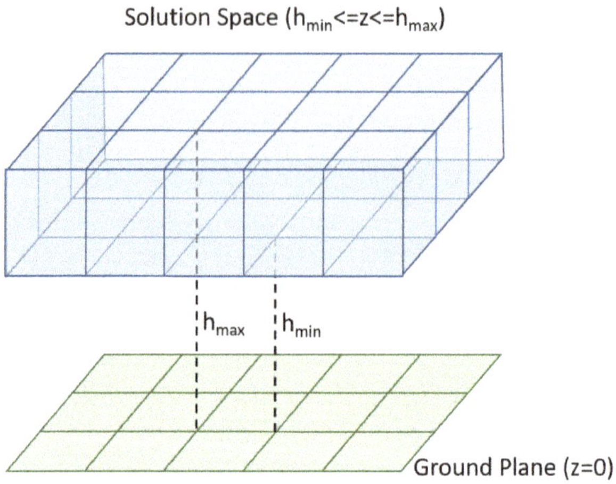

Figure 5.11 Illustration of bounded altitude constraints on a flat-Earth model.

where $R_{\text{eff}}(\phi)$ and $\mathbf{P}$ are defined similarly as in Section 5.1. These constraints can be generated by calling the function `boundedAlt` in the package `utils.constraints`.

5.2.2 Solvers

Just as the known altitude case represents an equality-constrained optimization problem, knowledge of altitude bounds represents an inequality-constrained optimization problem. For a discussion of how this is approached in convex optimization problems, see Chapter 11 of [1] on *interior point methods*. The focus of this chapter is on an approach referred to as the *barrier method*, although others are discussed.

We will discuss first how the general solutions can be modified to approach this problem and the limitations therein.

5.2.2.1 Generic Solvers

The modification of the brute force ML solution to the case of bounded altitude constraints is identical to the fixed altitude case, the solver is modified to set the likelihood of any points outside the solution space to $-\infty$, thereby suppressing any possibility of a solution outside the bounds. The same function achieves both fixed (equality) and bounded (inequality) constraints: `mlSolnConstrained`.

```
[x_est,A,x_grid] = mlSolnConstrained(x_tdoa,rho,C,x_ctr,search_size,epsilon,
    a,b,tol)
```

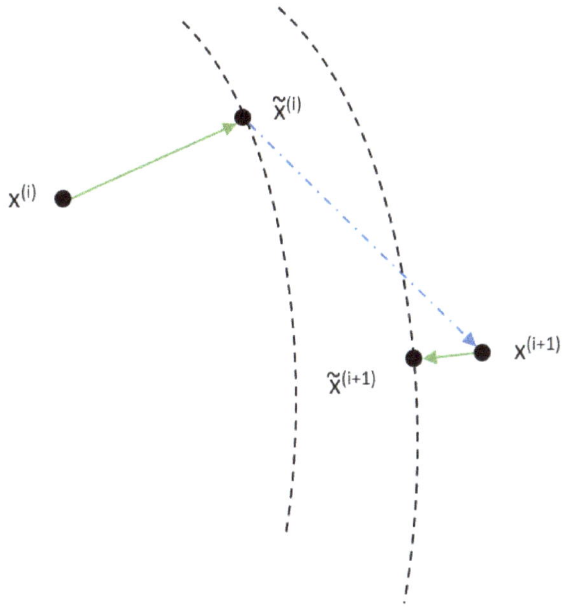

Figure 5.12 Illustration of the iterative-update approach to bounded altitude problems (in a 2-D example). Green arrows indicate the application of the constraints, while the blue dash-dot line is a single step of the unconstrained solver.

where `a` is a cell array of equality constraints, `b` is a cell array of inequality constraints, and `tol` is an optional tolerance for the equality constraints.

For the iterative approaches, we have also implemented the same iterative-adaptive approach where the unconstrained solver is used to update the geolocation estimate, and then the bounds are applied at each iteration. This behavior is illustrated for bounded altitude cases in Figure 5.12. In this case, we have the added step of determining which of the two inequality constraints (if any) are violated and need to be used to scale the result. This is implemented in the functions `lsSolnBounded` and `gdSolnBounded` in the `aoa`, `tdoa`, `fdoa`, and `hybrid` packages.

An example usage, in the `triang` package, is printed here. In this usage, all of the default parameters are left unspecified (the `alpha` and `beta` parameters that control the gradient descent algorithm, as well as the flags that control stopping upon convergence and the printing of interim results as the algorithm is executing). Usage for the other packages is similar; the difference between these methods and the original `gdSoln` function calls is the inclusion of the inequality constraints `b` after the initial position estimate.

```
[x,x_full] = gdSolnBounded(x_aoa,psi,C,x_init,b);
```

Example 5.4 Bounded Satellite Orbit Example

Consider three TDOA sensors, spaced at 50 km. The central sensor is located at 25°N, 15°W, 10 m alt, and the others are 50 km due east and due north, respectively (at the same altitude). There is a satellite located above the point 27°N, 13°W, at an altitude of 575 km.

Assume the altitude of the satellite is bounded between 500 and 600 km. Use iterative least squares to solver for the location of the satellite, both with and without the altitude constraints.

Solution: We begin by defining the sensors in ENU coordinates, then converting to ECEF, since we're dealing with a satellite in orbit and want to account for curvature of the Earth. The satellite is defined in Lat/Lon/Alt, and similarly converted to both ECEF for analysis and ENU for plotting.

```
ref_lla = [25, -15, 0];
x_aoa_e = [0, 50e3,    0];
x_aoa_n = [0,     0, 50e3];
x_aoa_u = [10,  10,   10];
[x,y,z] = utils.enu2ecef(x_aoa_e, x_aoa_n, x_aoa_u,...
                         ref_lla(1), ref_lla(2), ref_lla(3), 'deg','m');
x_aoa = [x(:), y(:), z(:)]'; % ECEF

n_aoa = size(x_aoa,2);

sat_lla = [27, -13, 575e3];
[x,y,z] = utils.lla2ecef(sat_lla(1), sat_lla(2), sat_lla(3), 'deg','m');
x_tgt = [x, y, z]'; % ECEF
[e,n,u] = utils.lla2enu(sat_lla(1), sat_lla(2), sat_lla(3), ...
                         ref_lla(1), ref_lla(2), ref_lla(3), 'deg', 'm');
x_tgt_enu = [e, n, u]';
```

We define the altitude constraint with a call to `boundedAlt`, with the `'ellipse'` flag, to implement the altitude constraint in (5.28).

```
alt_low = 500e3;
alt_high = 600e3;
b = utils.constraints.boundedAlt(alt_low, alt_high, 'ellipse');
```

Next, we generate a set of measurements, as we've done many times now, using the sensor covariance `C_aoa`.

```
err_aoa = 1*pi/180; % 1 deg
C_aoa = err_aoa^2*eye(2*n_aoa);
L = chol(C_aoa,'lower');
```

```
z = triang.measurement(x_aoa, x_tgt);
n = L*randn(2*n_aoa,1);
zeta = z+n;
```

Solving both problems (unconstrained and constrained) is a straightforward call to `gdSoln` and `gdSolnBounded`, respectively.

```
%% Solvers
[x,y,z] = utils.lla2ecef(ref_lla(1), ref_lla(2), 500e3);
x_init = [x, y, z]';

[x_gd, x_gd_full] = triang.gdSoln(x_aoa, zeta, C_aoa, x_init);
[x_gd_bnd, x_gd_bnd_full] = triang.gdSolnBounded(x_aoa,zeta,C_aoa,x_init,b);
```

The solutions are printed below, and the error for each result is printed. We see that the unconstrained solution is off by only .02 degrees latitude, but .64 deg longitude, and 155 km altitude, for a 3-D position error of just over 169 km. Clearly, most of the error is in the altitude dimension, which makes sense, given the geometry (three relatively closely spaced angle-only sensors geolocating a distance source). In the constrained case, we supply external knowledge that the altitude is bounded by 500 and 600 km. The result of this altitude bound is that the position estimate is much closer, off by only 97.35 km. This makes some sense, as the unconstrained solution is below the lower bound constraint, and so a constrained solution is going to stay at that bound.

```
Unconstrained Solution:  27.02 deg N, 13.64 deg W, 420.39 km
   Error: 169.21 km
Constrained Solution:    27.34 deg N, 13.43 deg W, 500.00 km
  Error:  97.35 km
```

We plot the sensors, target, and both the constrained and unconstrained solutions in Figure 5.13. Before plotting, the solutions were converted from ECEF to ENU for more convenient visualization. □

5.2.2.2 Barrier Method

Barrier methods refer to a class of problems that replace the firm inequality constraint (which ideally has no cost if it is satisfied, and infinite cost if it is violated) with a differentiable function that approximates the constraint. One such class of barriers is scaled logarithms [1]:

$$I(u) = -\left(\frac{1}{t}\right)\log(u) \tag{5.29}$$

where $t > 0$ is a sharpness parameter, and u is the output of an inequality constraint function (in our case, $b_n(\mathbf{x})$), which must be less than or equal to zero to satisfy the constraint. Figure 5.14 plots the ideal (dashed line) and several logarithmic barriers.

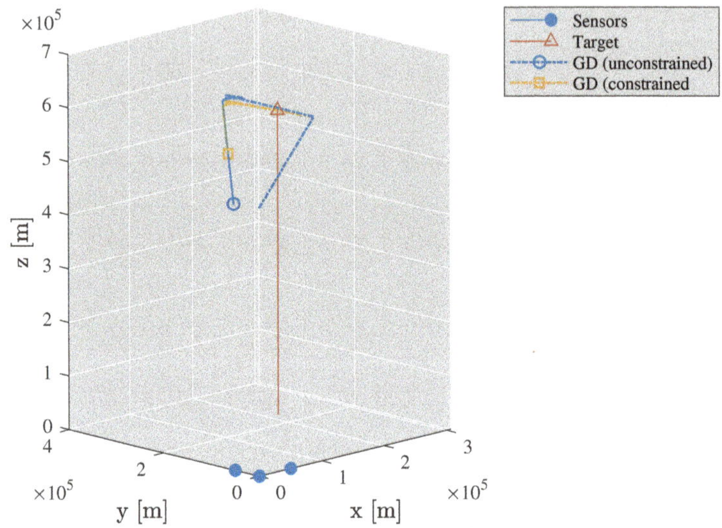

Figure 5.13 Illustration of unconstrained and fixed altitude solutions for Example 5.4.

Note that as $t \rightarrow \infty$, the barrier approaches the ideal shape.[13.] At values below the constraint, the shape of the barrier encourages solutions to stray from the inequality constraint but does not prevent them. Higher values of t will more accurately reflect the barrier but will be more likely to suffer numerical stability issues as the gradient is less helpful in discouraging solutions near the constraint.

For a discussion of how to solve the optimization problem, see Chapter 11 of [1].

This is incorporated into the optimization function via an approach similar to the Lagrange multipliers discussed in Section 5.1.1.

$$\tilde{\mathbf{x}} = \arg\left[\max_{\mathbf{x}} \ell(\zeta|\mathbf{x}) - (1/t)\sum_{n=0}^{N-1} \log\left(b_n(\mathbf{x})\right)\right] \tag{5.30}$$

There are many numerical methods presented in Chapter 11 of [1] that can be used to solve (5.30). For brevity, we do not implement any of them, but encourage interested readers to begin their research there.

13. Selection of the appropriate parameter t is a trade-off between convergence speed (better with lower t) and accuracy of the bound (better with higher t).

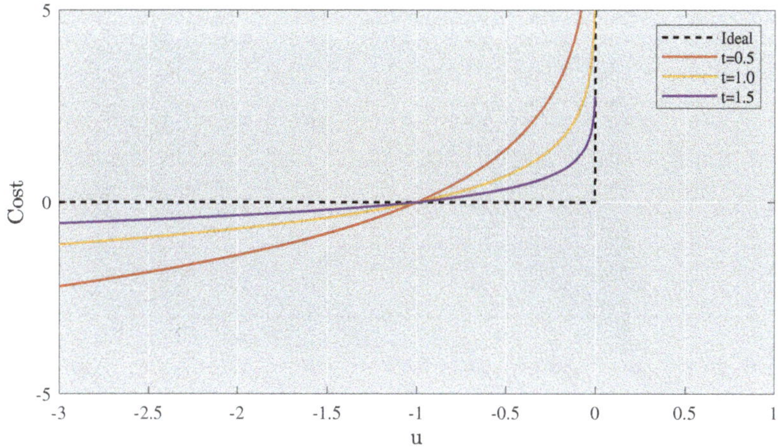

Figure 5.14 Illustration of logarithmic (solid) and ideal (dashed) barriers for inequality constraints ($u \leq 0$).

5.2.3 Performance Analysis

Unfortunately, the CRLB for inequality constrained problems cannot be solved in closed-form [19]. As such, performance can only be analyzed via Monte Carlo simulation. It is good to note, however, that inequality constraints have been shown to have no effect on the CRLB [3], [20]. And, thus, the standard CRLB from Section 2.4.5 can be utilized for known altitude bound conditions (provided the bounds are accurate).

5.3 GEOLOCATION WITH STATISTICAL PRIORS

The principal problem with hard geographic constraints, such as known altitude or geographic bounds, is that if the assumptions are incorrect, performance will suffer and the geolocation solution will almost certainly be more erroneous than expected. In cases where the target constraints are not guaranteed, then the uncertainty should be captured via statistical priors, represented with an a priori probability distribution function $f_{\mathbf{X}}(\mathbf{x})$. In this manner, the posterior likelihood function for any set of measurements ζ is given:

$$\ell(\zeta, \mathbf{x}) = \ell(\zeta|\mathbf{x}) f_{\mathbf{X}}(\mathbf{x}). \tag{5.31}$$

where $\ell(\zeta|\mathbf{x})$ is the *likelihood function*, which computes how likely it is that the measurement ζ would occur if the true target location were $\mathbf{x}$. The posterior likelihood function $\ell(\zeta, \mathbf{x})$ reflects the combination of the measurement likelihood function and the a priori probability distribution function $f_{\mathbf{X}}(\mathbf{x})$. This behavior is illustrated in Figure 5.15.

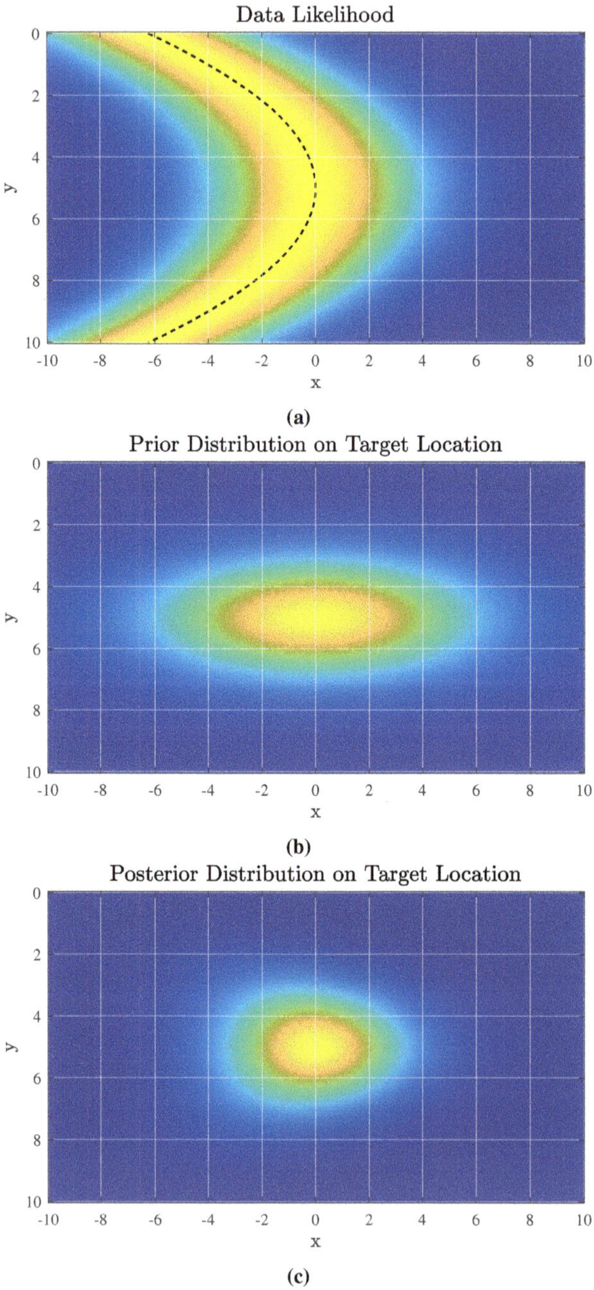

Figure 5.15 Example illustration of the use of prior information to adjust the likelihood. (a) Initial likelihood function, (b) prior statistical information, and (c) posterior product of the prior information and the likelihood function.

If we are using the log-likelihood (as we always do in this text), then the prior information must be similarly converted to log form and added to the log likelihood, rather than multiplied.

$$\ell(\zeta, \mathbf{x}) = \ell(\zeta|\mathbf{x}) + \log\left(f_{\mathbf{x}}(\mathbf{x})\right) \tag{5.32}$$

5.3.1 Example Priors

If a prior, or externally generated estimate, is available, it can be used to improve estimation performance, so long as its distribution is fully defined. This is most often expressed as a Gaussian distribution, with its mean as the provided estimate, and some covariance matrix defined by how confident or accurate that estimate is.

$$\mathbf{x} \sim \mathcal{N}\left(\widehat{\mathbf{x}}, \mathbf{C}_{\widehat{\mathbf{x}}}\right) \tag{5.33}$$

for some estimate $\widehat{\mathbf{x}}$ and error covariance $\mathbf{C}_{\widehat{\mathbf{x}}}$. Other distributions are equally valid, but we will stick with Gaussian distributions, for simplicity (particularly, since they are fully defined by just two parameters: the mean vector and covariance matrix).

5.3.2 Solving Geolocation with Priors

Any of the solutions previously proposed for geolocation (e.g., brute-force maximum likelihood estimation, or iterative least squares estimation) apply just as well to geolocation with priors, as to the original problems as posed. The only potential drawback is that the iterative least square and convex solvers previously derived all assume that the problem is *locally* convex, so that the gradient of the likelihood function can be used to search for the optimal solution, whereas the priors may make the problem less convex. This is not an issue for priors that are defined with a smooth distribution, such as a Gaussian prior.[14]

This is implemented by defining the prior as a function that accepts target locations and returns a probability, which is then passed to the modified ML solver: `mlSolnPrior`, in the relevant directory. The iterative solvers for gradient descent and iterative least squares do not inherently consider the likelihood of a given set of measurements, and so the modification of these algorithms to include statistical priors is more involved.[15]

For ML with priors, the log of the prior is added to the log-likelihood function, according to (5.32).[16] For example, the versions in the `tdoa` package use the following two lines to generate and then modify the function handle for the likelihood function.

14. For a discussion of operations that preserve convexity, see Section 3.2 of [1].
15. See, for example, Section 7.1.2 of [1] for a discussion of how the prior distribution affects the convex optimization problem.
16. While the formulation in (5.32) is mathematically complete, we have chosen to include an optional parameter $\lambda \in [0, 1]$ that is used to provide a linear weight between the likelihood and the prior ($\lambda = 1$). The default value of $\lambda = .5$ is mathematically equivalent to the formulation in (5.32), and any deviation represents more heavily weighting either the measurements ($\lambda < .5$) or the prior ($\lambda > .5$).

```
ell = tdoa.likelihood( . . . );
ell_posterior = @(x) ell(x) + log(prior(x));
```

Example 5.5 External Estimate informing Ground-Based TDOA

Consider a ground-based TDOA system with four receivers, arrayed in a Y-configuration, with one sensor at the center, and the others 10 km away at due north, 30° south of west, and 30° south of east (in other words, at bearings of 0°, 120°, and 240° east of north). The timing error of each receiver is 1 μs. An emitter is located 100 km away, due east, at an altitude of 20 kft. Assume, for this problem, that the Earth is spherical.

An external sensor has produced an estimate of the target's position at $[95,\ 0,\ 10]$ km in local ENU coordinates, with a covariance matrix

$$\mathbf{C} = \begin{bmatrix} 5 & 1 & 0 \\ 1 & 50 & 0 \\ 0 & 0 & 10 \end{bmatrix} * 10^6 \text{ m}^2 \tag{5.34}$$

Construct the Gaussian prior, and use it to solve the geolocation problem with random sensor noise using the maximum likelihood solver, using a grid that extends ± 50 km east and north, and is centered on the target's true up coordinate, with a spacing of 250 m per sample point. Compare to the geolocation result without the prior.

Solution: We define the TDOA sensor positions in a straightforward manner, using some trigonometry, so that we can easily increase the number of receivers in the ring around the central reference. Since the default behavior is for the nth receiver to be the common reference, we'll put that one at the end. The process for defining the target position is slightly more involved. To account for the Earth's curvature at long range we first define the position in *ground units*; that is in meters east and north along the Earth's surface, and then the altitude perpendicular. We can use these directly as ENU, but then there will be a significant difference between the stated "up" coordinate and the altitude of the specified position above the Earth's surface. To account for a spherical Earth, we use the utility `correctENU`. This utility converts the specified coordinates (in *east/north along the Earth's surface, and local altitude*) to Cartesian ENU. In our case, the net result is that while our target is 12.1 km (40 kft) above the Earth's surface, it's *up* coordinate is actually 11.41 km.

```
baseline = 10e3;
n_tdoa = 4;
tdoa_ang = pi/6 + 2*pi/3 *(0:n_tdoa-2);
x_tdoa = baseline * [cos(tdoa_ang), 0; ...
                     sin(tdoa_ang), 0;
                     0, 0, 0, 0]; % ENU
```

```
% Define target coordinates
tgt_rng = 100e3;
tgt_alt = 40e3*unitsratio('m','ft');
x_tgt_g = [tgt_rng; 0; tgt_alt]; % spherical Earth coors (East,North,Alt)
[e, n, u] = utils.correctENU(x_tgt_g(1), x_tgt_g(2), x_tgt_g(3));
x_tgt = [e; n; u]; % ENU
```

To generate the prior, we first input the prior position estimate and covariance matrix, and then use them to generate a function handle that calls `mvnpdf` with them as the mean and covariance. Note that `mvnpdf` works on rows, whereas our input will be specified as columns, so we must transpose both `x` and `x_prior`. Just as with the down-range target position, we'll convert our prior to ENU to correct for the Earth's curvature.

```
x_prior_g = [95; 10; 10]*1e3;
[e, n, u] = utils.correctENU(x_prior_g(1), x_prior_g(2), x_prior_g(3));
x_prior = [e; n; u];
C_prior = [20, 5, 0; 5, 10, 0; 0, 0, 10]*1e5;

prior = @(x) mvnpdf(x', x_prior', C_prior);
```

The process of generating a measurement proceeds as normal.

```
z = tdoa.measurement(x_tdoa, x_tgt);

time_err = 3e-7;
C_roa = (utils.constants.c*time_err)^2 * eye(n_tdoa);
C_rdoa = utils.resampleCovMtx(C_roa,[]);
L = chol(C_rdoa,'lower');

n = U*randn(size(z));

zeta = z+n;
```

What remains is the straightforward process of defining a search grid, and then executing the maximum likelihood geolocation both with and without the prior.

```
x_ctr = x_tgt;
grid_size = [50e3, 50e3, 0];
epsilon = 250;

[x_ml, A, x_grid] = tdoa.mlSoln(x_tdoa, zeta, C_roa, x_ctr, grid_size,
    epsilon);
[x_ml_p, A_p, ~] = tdoa.mlSolnPrior(x_tdoa, zeta, C_roa, prior, x_ctr,
    grid_size, epsilon);
```

The geolocation estimates are printed here and plotted in Figure 5.16. We note that the original case, without a prior, in Figure 5.16(a), shows an expected result, good angular geolocation, and relatively poor down-range geolocation from the TDOA sensors. The solution is approximately 35 km away from the true target position, which makes sense given the 500-ns timing error that we've assumed. In the case with the prior, however, we see in Figure 5.16(b) the white dashed line, indicating the ellipse drawn for a 90% confidence, which provides some range information. Even though the prior that we've given is approximately 12 km from the true position, the error in the resulting estimate is reduced to 11 km, because the prior was given equal weight to the measurements, allowing us to leverage the new measurements to improve the north and up coordinates, while the prior improved our ability to estimate the east coordinate (slightly). Note that the true target position is outside the 90% confidence ellipse provided by the prior, so in this case the statistical prior was not accurate, yet still provided some benefit.

```
Solution w/o prior: 64.44 km, -1.75 km, 11.41 km
        Error: 35.79 km
Solution w/prior:   89.44 km, -0.75 km, 11.41 km
        Error: 10.78 km
```

□

5.3.3 Performance Analysis

To analyze performance, we'll take the CRLB and show how it can be modified to include the prior distribution on **x**. We begin with the stochastic definition of the Fisher information matrix [21], applied to the log-likelihood function $\ell(\zeta|\mathbf{x})$:

$$F_{i,j}(\zeta|\mathbf{x}) = E\left[\frac{\partial^2}{\partial x_i \partial x_j}\ell(\zeta|\mathbf{x})\right] \tag{5.35}$$

In our case, we simply apply this same formulation to the posterior likelihood function and simplify to its components

$$F_{i,j}\left(\mathbf{x},\zeta\right) = E\left[\frac{\partial^2}{\partial x_i \partial x_j}\left(\ell(\zeta|\mathbf{z}) + \log\left(f_{\mathbf{x}}(\mathbf{x})\right)\right)\right] \tag{5.36}$$

$$= E\left[\frac{\partial^2}{\partial x_i \partial x_j}\ell(\zeta|\mathbf{z})\right] + E\left[\frac{\partial^2}{\partial x_i \partial x_j}\log\left(f_{\mathbf{x}}(\mathbf{x})\right)\right] \tag{5.37}$$

$$= F_{i,j}\left(\zeta|\mathbf{x}\right) + F_{i,j}(\mathbf{x}) \tag{5.38}$$

where $F_{i,j}(\zeta|\mathbf{x})$ is the standard Fisher information matrix based on the likelihood of the received data conditioned on the target position **x**, defined in (5.35), and $F_{i,j}\left(\mathbf{x}\right)$ is the Fisher information matrix of the prior. So, this must be solved for whatever prior is provided. In the case of a Gaussian prior with mean $\widehat{\mathbf{x}}$ (that is assumed to be an unbiased

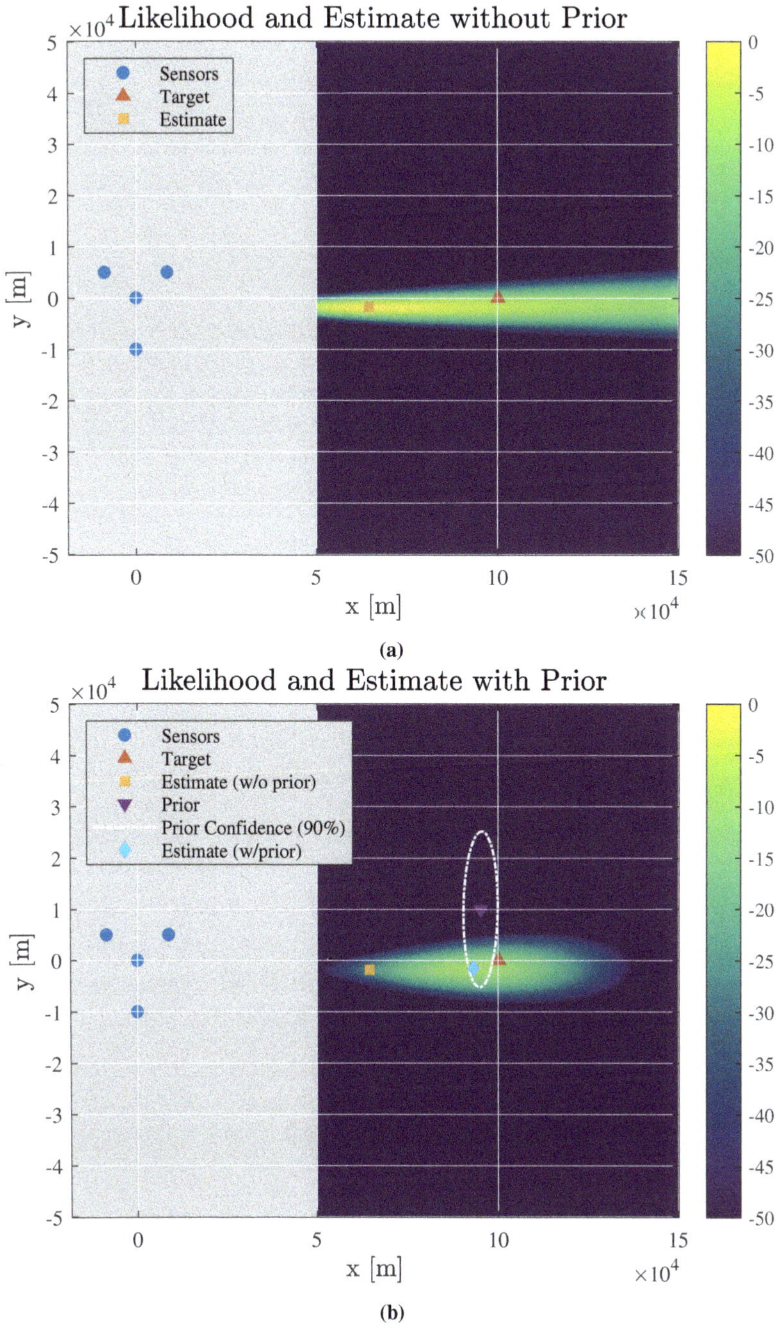

Figure 5.16 Illustration of the ML geolocation result for Example 5.5, (a) without the prior distribution, and (b) with the prior distribution. The latter also shows the expectation from the prior (purple inverted triangle) and its associated 90% confidence error ellipse (white dashed line).

estimate of $\mathbf{x}$), and error covariance $\mathbf{C}_{\widehat{\mathbf{x}}}$, the Fisher information matrix of the prior is:

$$\mathbf{F}(\mathbf{x}) = \mathbf{C}_{\widehat{\mathbf{x}}}^{-1} \tag{5.39}$$

and, thus, the CRLB with a prior distribution on the positions is:

$$\mathbf{C}_{\mathbf{x}} \geq \left[\mathbf{F}(\zeta|\mathbf{x}) + \mathbf{C}_{\widehat{\mathbf{x}}}^{-1}\right]^{-1} \tag{5.40}$$

$$= \left[\mathbf{J}_z^T(\mathbf{x})\mathbf{C}_\zeta^{-1}\mathbf{J}_z(\mathbf{x}) + \mathbf{C}_{\widehat{\mathbf{x}}}^{-1}\right]^{-1} \tag{5.41}$$

To implement this, we've built a utility function `makePrior` that will return two function handles; one for the prior distribution ($f_{\mathbf{x}}(\mathbf{x})$), and one for the Fisher information matrix ($\mathbf{F}(\mathbf{x})$).[17] The latter can then be passed into a modified form of the CRLB, called `computeCRLBpriors` that will accept all the of the same inputs as the matching function `computeCRLB`, as well as the function handle to the FIM of the prior, and will compute the posterior FIM from (5.41).

```
[prior, fim_prior] = utils.makePrior('gaussian', prior_mean,
    prior_covariance);
crlb = tdoa.computeCRLBpriors({ typical crlb params }, fim_prior);
```

5.4 PROBLEM SET

5.1 Derive the partial derivatives $\frac{\partial a(\mathbf{x})}{\partial x}$, $\frac{\partial a(\mathbf{x})}{\partial y}$, and $\frac{\partial a(\mathbf{x})}{\partial z}$ in (5.16–5.18).

5.2 Repeat Example 5.2, but using the iterative least square solvers, instead of gradient descent. Comment on the differences with and without the target altitude constraint, and compare with the gradient descent case.

5.3 Consider Example 5.3, and generate noisy measurements for 1,000 Monte Carlo trials. Implement an iterative least square solver and compare the RMSE of the geolocation estimates to the CRLB from Example 5.3 (both with and without knowledge of the target altitude).

5.4 Repeat Example 5.3 with FDOA receivers instead of TDOA, and with 100-Hz frequency error, 3-GHz carrier frequency, and sensor velocity of 100 m/s at a heading of 45°E of N.

5.5 Consider Example 5.4, generate noisy measurements and solution estimates for 1,000 Monte Carlo trials and compute the RMSE. Compute the unbiased CRLB and compare to the performance of the bounded and unbounded solvers.

17. See `help utils.makePrior` for usage instructions. Currently only Gaussian and uniform priors are implemented.

5.6 Define a fixed constraint in ENU coordinates for targets at $E = 100$ km.

5.7 Define a bounded constraint in ENU coordinates for targets between 100 km and 200 km in the north dimension.

5.8 Define a fixed altitude constraint in ECEF for a satellite at 1,500 km.

References

[1] S. Boyd and L. Vandenberghe, *Convex Optimization*. Cambridge, UK: Cambridge University Press, 2004.

[2] J. Gorman and A. Hero, "Lower bounds on parametric estimators with constraints," *Proc. Fourth Annual ASSP Workshop on Spectrum Estimation and Modeling*, pp. 223–228, 1988. DOI: 10.1109/SPECT.1988.206196.

[3] J. Gorman and A. Hero, "Lower bounds for parametric estimation with constraints," *IEEE Transactions on Information Theory*, vol. 36, no. 6, pp. 1285–1301, 1990. DOI: 10.1109/18.59929.

[4] Wikipedia, *Automated identification system*. [Online]. Available: https://en.wikipedia.org/wiki/Automatic_identification_system#cite_note-businesswire.com-16.

[5] B. Wire, *Orbcomm announces launch window for second og2 mission*, 2015. [Online]. Available: https://www.businesswire.com/news/home/20151016005656/en/ORBCOMM-Announces-Launch-Window-OG2-Mission.

[6] K. Ho and Y. Chan, "Geolocation of a known altitude object from TDOA and FDOA measurements," *IEEE Transactions on Aerospace and Electronic Systems*, vol. 33, no. 3, pp. 770–783, 1997. DOI: 10.1109/7.599239.

[7] G. Niezgoda and K. Ho, "Geolocalization by combined range difference and range rate difference measurements," *Proceedings of IEEE International Conference on Acoustics, Speech and Signal Processing (ICASSP '94)*, vol. 2, pp. 357–360, 1994. DOI: 10.1109/ICASSP.1994.389616.

[8] N. A. O'Donoughue, *Emitter Detection and Geolocation for Electronic Warfare*. Norwood, MA: Artech House, 2019.

[9] Y. Z. Bin, W. Lei, C. P. Qun, and L. A. Nan, "Passive satellite localization using TDOA/FDOA/AOA measurements," in *IEEE Conference Anthology*, 2013, pp. 1–5. DOI: 10.1109/ANTHOLOGY.2013.6784815.

[10] Y. Cao, L. Peng, J. Li, L. Yang, and F. Guo, "A new iterative algorithm for geolocating a known altitude target using TDOA and FDOA measurements in the presence of satellite location uncertainty," *Chinese Journal of Aeronautics*, vol. 28, no. 5, pp. 1510–1518, 2015.

[11] T. Marzetta, "A simple derivation of the constrained multiple parameter cramer-rao bound," *IEEE Transactions on Signal Processing*, vol. 41, no. 6, pp. 2247–2249, 1993. DOI: 10.1109/78.218151.

[12] P. Vergez, L. Sauter, and S. Dahlke, "An improved Kalman filter for satellite orbit predictions," *The Journal of the Astronautical Sciences*, vol. 52, no. 3, pp. 359–380, 2004.

[13] M. F. Storz, B. R. Bowman, M. J. I. Branson, S. J. Casali, and W. K. Tobiska, "High accuracy satellite drag model (HASDM)," *Advances in Space Research*, vol. 36, no. 12, pp. 2497–2505, 2005, ISSN: 0273-1177. DOI: https://doi.org/10.1016/j.asr.2004.02.020.

[14] J. Sharma, G. H. Stokes, C. von Braun, G. Zollinger, and A. J. Wiseman, "Toward operational space-based space surveillance," *Lincoln Laboratory Journal*, vol. 13, no. 2, pp. 309–334, 2002.

[15] J. V. Eshbaugh, R. L. Morrison, E. W. Hoen, T. C. Hiett, and G. R. Benitz, "Husir signal processing," *Lincoln Laboratory Journal*, vol. 21, no. 1, pp. 115–134, 2014.

[16] A. M. Wathen, "The Miracle of Operation Iraqi Freedom Airspace Management," *Air & Space Power Chronicles–Chronicles Online Journal*, vol. 4, 2005.

[17] A. Wathen, "Joint airspace management and deconfliction: A chance to trade in a stovepipe for network-centric warfare," *Air & Space Power Journal*, vol. 20, no. 3, p. 26, 2006.

[18] D. Griffith, G. K. Wilson-Smith, M. Ohmer, *et al.*, "Coalition airspace management and deconfliction," Air Force Research Laboratory, Rome, NY, Tech. Rep., 2008.

[19] A. Benavoli, A. Farina, and L. Ortenzi, "MLE in presence of equality and inequality nonlinear constraints for the ballistic target problem," in *2008 IEEE Radar Conference*, 2008, pp. 1–6. DOI: 10.1109/RADAR.2008.4721042.

[20] Z. Ben-Haim and Y. C. Eldar, "On the Constrained Cramer–Rao Bound With a Singular Fisher Information Matrix," *IEEE Signal Processing Letters*, vol. 16, no. 6, pp. 453–456, 2009. DOI: 10.1109/LSP.2009.2016831.

[21] L. L. Scharf, *Statistical Signal Processing*. Reading, MA: Addison-Wesley, 1991.

Chapter 6

Geolocation with Sensor Uncertainties

Just as it is reasonable in an introductory physics course to ignore complex concepts, such as wind resistance, in the interest of understanding basic concepts, it was reasonable that we assumed perfect knowledge of sensor parameters in our first text [1]. This, however, is often unrealistic. In large, dedicated platforms, errors in the position, orientation, and velocity of a platform may be manageable, but for small systems, even small satellites, sensor uncertainties can become the driving factor in geolocation performance [2], [3]. In the age of global satellite navigation systems, we can often get away with ignoring these errors, but they are present nonetheless, and military systems must be prepared to operate in a contested electromagnetic spectrum, where such satellite navigation systems may be degraded or unavailable. As an example, the technology to spoof GPS systems has been proven in university settings, and has a shockingly low barrier to entry [4]–[6]. For all these reasons, it is wise to understand the impacts of sensor uncertainties on performance.

In this chapter, we will continue to assume that sensor *performance* is well understood. That is, we can accurately predict how well our timing estimates or angle-of-arrival estimates are or will be. We are instead focused on unknown measurement biases and variances in the sensor's position and velocity.

6.1 FORMULATION

To begin, we'll introduce how the errors will be modeled, and reformulate measurement, likelihood, gradients, and the Fisher information matrices used to predict geolocation performance to account for these uncertainties. We consider, broadly, five types of sensor uncertainty, listed in Table 6.1.

These errors take two general forms: (a) measurement biases (orientation, timing, and frequency errors), and (b) geolocation uncertainties (position and velocity errors). The former are generally taken as an unknown deterministic bias, while the latter are generally taken as a random variable. The principal reason for treating measurements

Table 6.1

Sensor Uncertainties

Error Type	Formulation	Size
Sensor position errors	$\overline{\mathbf{x}} \sim \mathcal{N}(0, \mathbf{C}_\mathbf{x})$	$N_d(N_a + N_t + N_f) \times 1$
Sensor velocity errors	$\overline{\mathbf{v}} \sim \mathcal{N}(0, \mathbf{C}_\mathbf{v})$	$N_d N_f \times 1$
Sensor orientation errors	$\overline{\boldsymbol{\psi}}, \overline{\boldsymbol{\phi}}$	$N_a \times 1$
Sensor range errors†	$\overline{\mathbf{r}}$	$M_t \times 1$
Sensor range-rate errors†	$\overline{\dot{\mathbf{r}}}$	$M_f \times 1$

N_d = number of spatial dimensions (2 or 3),
N_a = number of AOA sensors,
N_t = number of TDOA sensors,
N_f = number of FDOA sensors,
M_t = number of TDOA measurements,
M_f = number of FDOA measurements,
†The TDOA and FDOA biases are most easily thought of a time and frequency bias terms, but the notation and derivation is greatly simplified if we express them as range and range-rate biases.

as deterministic unknowns is that random measurement errors are already considered in the general formulations from Chapter 2 and that fixed bias terms are known to occur in a variety of scenarios.[1]

For position and velocity errors, a random error model is most appropriate largely due to the dynamic nature of the error sources, which have not yet been considered in our formulations (we assumed perfect knowledge of sensor position and velocity). Because of these differing models, we'll rely on the hybrid CRLB as opposed to the traditional CRLB (which is defined for unknown deterministic parameters, rather than random variables) [9].

Generally, let us denote $\boldsymbol{\alpha}$ as the vector of all sensor uncertainties. This will be a subset of the orientation $\left(\overline{\boldsymbol{\psi}}, \overline{\boldsymbol{\phi}}\right)$, range $\left(\overline{\mathbf{r}}\right)$, and range-rate $\left(\overline{\dot{\mathbf{r}}}\right)$ biases, noted in the three lower rows of Table 6.1.[2] Recall that our measurement vector $\boldsymbol{\zeta}$ is assumed to have a Gaussian distribution with mean value $\mathbf{z}(\mathbf{x})$ and covariance $\mathbf{C}_\zeta$ for some target located at $\mathbf{x}$. To first account for sensor biases $\boldsymbol{\alpha}$, we introduce a slight change in notation. Let

1. For example, in addition to more obvious bias sources, such as compass calibration errors for AOA, and clock or local oscillator drift between sensors for TDOA and FDOA, more complex phenomena such as array calibration errors [7] and mutual coupling [8] can cause biases in the received signals used to estimate angle, time, or frequency of arrival. Many of these biases are indeed time varying, but we will assume that they are stationary over the course of a measurement interval for this work.
2. It is more convenient to express TDOA measurement biases in terms of range bias rather than time bias (with $\overline{\mathbf{r}} = c\overline{\mathbf{t}}$ for time bias $\overline{\mathbf{t}}$), and similarly for FDOA in terms of range-rate bias rather than frequency bias (with $\overline{\dot{\mathbf{r}}} = c\overline{\mathbf{f}}/f_0$ for frequency bias $\overline{\mathbf{f}}$ and carrier frequency f_0).

ζ have mean value $\mathbf{z}(\mathbf{x}, \boldsymbol{\alpha})$, defined as

$$\mathbf{z}(\mathbf{x}, \boldsymbol{\alpha}) = \mathbf{z}(\mathbf{x}) + \mathbf{A}\boldsymbol{\alpha} \tag{6.1}$$

$$\mathbf{A} = \text{diag}\left\{ \underbrace{\mathbf{I}_{N_a}}_{\text{Azimuth}}, \underbrace{\mathbf{I}_{N_a}}_{\text{Elevation}}, \underbrace{\mathbf{A}_t}_{\text{TDOA}}, \underbrace{\mathbf{A}_f}_{\text{FDOA}} \right\} \tag{6.2}$$

where $\mathbf{A}$ is a mapping matrix that translates the sensor measurement errors to the measurement space used by $\mathbf{z}$ and ζ. For TDOA and FDOA, that means translating from the N_t or N_f sensors to the M_t or M_f measurements, respectively. The mapping matrices $\mathbf{A}_t$ and $\mathbf{A}_f$ achieve this translation, and are defined:

$$[\mathbf{A}_t]_{n,m} = \begin{cases} 1 & n = \mathbf{a}_{t,m} \\ -1 & n = \mathbf{b}_{t,m} \\ 0 & \text{else} \end{cases} \tag{6.3}$$

$$[\mathbf{A}_f]_{n,m} = \begin{cases} 1 & n = \mathbf{a}_{f,m} \\ -1 & n = \mathbf{b}_{f,m} \\ 0 & \text{else} \end{cases} \tag{6.4}$$

where $\mathbf{a}_t$ and $\mathbf{b}_t$ are the test and reference indices for TDOA, and $\mathbf{a}_f$ and $\mathbf{b}_f$ are the test and reference indices for FDOA, as defined in Chapter 3.

We express the log-likelihood of ζ with sensor bias terms $\boldsymbol{\alpha}$:

$$\ell\left(\zeta|\mathbf{x}, \boldsymbol{\alpha}\right) = -\frac{1}{2}\left(\zeta - \mathbf{z}(\mathbf{x}, \boldsymbol{\alpha})\right)^T \mathbf{C}_\zeta^{-1}\left(\zeta - \mathbf{z}(\mathbf{x}, \boldsymbol{\alpha})\right) \tag{6.5}$$

The next modification of the likelihood function is to consider errors in sensor position ($\overline{\mathbf{x}}$) and velocity ($\overline{\mathbf{v}}$), noted generally with $\boldsymbol{\beta}$. The vector is modeled as a Gaussian random variable with mean $\boldsymbol{\beta}^0$ and covariance $\mathbf{C}_\beta$, where $\boldsymbol{\beta}^0$ is the vector of reported sensor positions and velocities. This is included in the likelihood function with a secondary term. We assume that the random sensor position errors are independent of the measurement errors and biases.

$$\begin{aligned}\ell\left(\zeta|\mathbf{x}, \boldsymbol{\alpha}, \boldsymbol{\beta}\right) = &-\frac{1}{2}\left(\zeta - \mathbf{z}(\mathbf{x}, \boldsymbol{\alpha}, \boldsymbol{\beta})\right)^T \mathbf{C}_\zeta^{-1}\left(\zeta - \mathbf{z}(\mathbf{x}, \boldsymbol{\alpha}, \boldsymbol{\beta})\right) \\ &-\frac{1}{2}\left(\boldsymbol{\beta} - \boldsymbol{\beta}^0\right)^T \mathbf{C}_\beta\left(\boldsymbol{\beta} - \boldsymbol{\beta}^0\right)\end{aligned} \tag{6.6}$$

We implement (6.6) in the function `likelihoodUnc` in each of the `triang`, `tdoa`, `fdoa`, and `hybrid` packages.

Note that the measurement vector $\mathbf{z}$ now explicitly considers both measurement biases $\boldsymbol{\alpha}$ and sensor position errors $\boldsymbol{\beta}$. The equation for $\mathbf{z}$ follows the same form as those in Chapter 2, but with additional terms:

$$\mathbf{z}(\mathbf{x}, \boldsymbol{\alpha}, \boldsymbol{\beta}) = \begin{bmatrix} \boldsymbol{\psi}(\mathbf{x}, \boldsymbol{\alpha}, \boldsymbol{\beta}) \\ \boldsymbol{\phi}(\mathbf{x}, \boldsymbol{\alpha}, \boldsymbol{\beta}) \\ \mathbf{r}(\mathbf{x}, \boldsymbol{\alpha}, \boldsymbol{\beta}) \\ \dot{\mathbf{r}}(\mathbf{x}, \boldsymbol{\alpha}, \boldsymbol{\beta}) \end{bmatrix} \tag{6.7}$$

$$\psi_m(\mathbf{x}, \boldsymbol{\alpha}, \boldsymbol{\beta}) = \tan^{-1}\left(\frac{y - \overline{y}_m}{x - \overline{x}_m}\right) + \overline{\boldsymbol{\psi}}_m \tag{6.8}$$

$$\phi_m(\mathbf{x}, \boldsymbol{\alpha}, \boldsymbol{\beta}) = \tan^{-1}\left(\frac{z - \overline{z}_m}{R_{g,m}(\mathbf{x})}\right) + \overline{\boldsymbol{\phi}}_m \tag{6.9}$$

$$R_m(\mathbf{x}, \boldsymbol{\alpha}, \boldsymbol{\beta}) = \left\|\mathbf{x} - \overline{\mathbf{x}}_{a_{t,m}}\right\| - \left\|\mathbf{x} - \overline{\mathbf{x}}_{b_{t,m}}\right\| + \overline{\mathbf{r}}_{a_{t,m}} - \overline{\mathbf{r}}_{b_{t,m}} \tag{6.10}$$

$$\dot{R}_m(\mathbf{x}, \boldsymbol{\alpha}, \boldsymbol{\beta}) = \overline{\mathbf{v}}_{a_{f,m}}^T \mathbf{u}_{a_{f,m}}(\mathbf{x}, \boldsymbol{\beta}) - \overline{\mathbf{v}}_{b_{f,m}}^T \mathbf{u}_{b_{f,m}}(\mathbf{x}, \boldsymbol{\beta}) + \overline{\dot{\mathbf{r}}}_{a_{f,m}} - \overline{\dot{\mathbf{r}}}_{b_{f,m}} \tag{6.11}$$

where the vector $\mathbf{u}_n(\mathbf{x}, \boldsymbol{\beta})$ is the unit vector pointing from the nth sensor (at the estimated position) to the target at $\mathbf{x}$:

$$\mathbf{u}_n(\mathbf{x}, \boldsymbol{\beta}) = \frac{\mathbf{x} - \overline{\mathbf{x}}_n}{\left\|\mathbf{x} - \overline{\mathbf{x}}_n\right\|} \tag{6.12}$$

and $R_{g,n}(\mathbf{x})$ is the range between the nth AOA sensor and a target at $\mathbf{x}$ *in the x-y plane* (sometimes referred to as the *ground plane*):

$$R_{g,n}(\mathbf{x}) = \sqrt{(x - \overline{x}_n)^2 + (y - \overline{y}_n)^2} \tag{6.13}$$

The original `measurement` functions of the `triang`, `fdoa`, `tdoa`, and `hybrid` packages have been augmented to take an optional bias term $\boldsymbol{\alpha}$, but the sensor position term $\boldsymbol{\beta}$ has always been an explicit definition of the `measurement` MATLAB® functions, and so no change is necessary.

Example 6.1 Direction Finding with Bias

Consider a 3-sensor DF problem in 2-D. Let the sensors be located at (2, 2) km, (2, −1) km, and (0, 0) km. A single transmitter is located at (3, 5) km. Generate the AOA measurements for a case where the bias on the three sensors is $\boldsymbol{\alpha} = [5°, 10°, -5°]$. Plot the LOBs for an ideal case (no bias) and for the case with bias.

Solution: First, we define the sensor and target positions.

```
x_aoa = [2, 2, 0; 2, -1, 0];
n_aoa = size(x_aoa,2);

x_tgt = [5; 3];
```

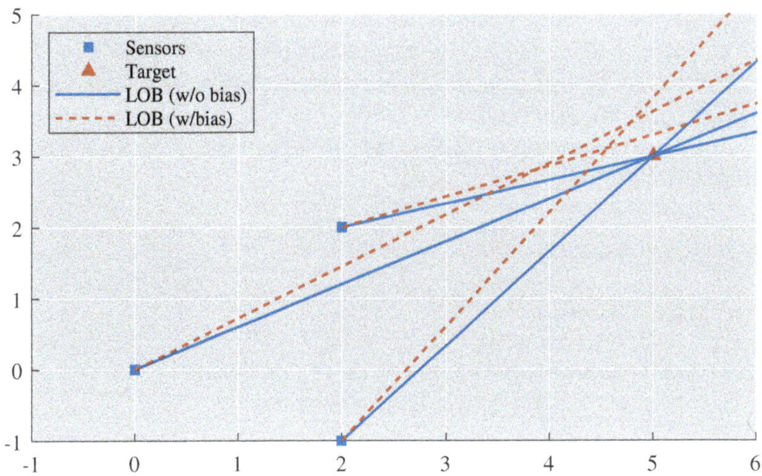

Figure 6.1 DF case with bias, from Example 6.1.

To generate the measurements, we call `triang.measurements` with the sensor and target positions. If we provide the bias term `alpha`, then it will be applied to the AOA measurements.

```
alpha = [5, 10, -5]*pi/180; % AOA bias
psi = triang.measurement(x_aoa, x_tgt);
psi_bias = triang.measurement(x_aoa, x_tgt, false, alpha);
```

To plot the LOBs, we made use of the command `triang.drawLOB` to generate the start and end point of each LOB conveniently. These are shown in Figure 6.1, where we can see that without bias, the three LOBs clearly intersect at the target's position, while bias shifts their intersection and, even without additional measurement noise, prevents all three LOBs from intersecting at the same point. □

Example 6.2 TDOA with Position Uncertainty

Consider a 3-sensor TDOA system arranged in a square with sensors at (0, 2) m, (2, -1) m, and (0, 0) m. The final sensor is a common reference for the other two, generating two TDOA measurements. Assume that the position of each sensor is independent and identically distributed with expectation equal to the nominal positions above and covariance of $\mathbf{C} = \sigma^2\mathbf{I}$, $\sigma^2 = 0.1m^2$. Generate a random position for each of the four sensors. Generate noise-free TDOA measurements for a target at (6, 3), using both the nominal and the noisy sensor positions, and compare them. Plot the isochrones using the measurements generated with the true (random) positions, for both sets of sensor positions.

Solution: First, we set up the nominal (reported) sensor positions, then generate a set of random positions that are dictated by the position covariance matrix C_pos. Since the sensor positions are IID, we can create that covariance matrix as a simple diagonal with 0.1 on each entry. We then generate 2*n_tdoa random perturbations (x and y for each sensor) distributed according to C_pos, then reshape them to match x_tdoa and add them to generate the true sensor positions.

```
x_tdoa = [0, 2, 0; 2, -1, 0]; % nominal position
n_tdoa = size(x_tdoa,2);

C_pos = .1*eye(2*n_tdoa);
L = chol(C_pos_full,'lower');
perturbations = reshape(L*randn(2*n_tdoa,1),2,n_tdoa);
beta = x_tdoa + perturbations; % true position
```

We generate both sets of measurements with tdoa.measurement, but for the true sensor positions we replace x_tdoa (the nominal positions) with beta (the true positions). We see that the difference is on the order of .2–.4 m in the reported RDOA for each measurement. To put this into perspective, we plot the isochrones in Figure 6.2. The squares and the solid lines plot the nominal sensor positions, and the isochrones that would be generated if we used these with data collected from the true sensor positions. The circles and the dashed lines plot the true sensor positions, and the isochrones that would be generated if we knew the true sensor positions. In this example, there is a significant difference between the true isochrones (dashed lines) and those that would be reported if we used the nominal sensor positions.

```
x_tgt = [6; 3];

zeta = tdoa.measurement(x_tdoa, x_tgt, n_tdoa);
zeta_unc = tdoa.measurement(beta, x_tgt, n_tdoa);
```

```
Measurements from sensors 1-2 (w.r.t sensor 0):
Nominal Positions: -0.63 m, -0.31 m
Random Positions:  -0.97 m, -0.14 m
```

□

6.1.1 Gradient of the Measurement Function with Sensor Uncertainties

As we have shown before, both the solutions and the performance analysis will depend heavily on the gradient of the measurement function $\mathbf{z}(\mathbf{x}, \boldsymbol{\alpha}, \boldsymbol{\beta})$ with respect to each of the unknown parameters:

$$\boldsymbol{\vartheta} = \begin{bmatrix} \mathbf{x} \\ \boldsymbol{\alpha} \\ \boldsymbol{\beta} \end{bmatrix} \tag{6.14}$$

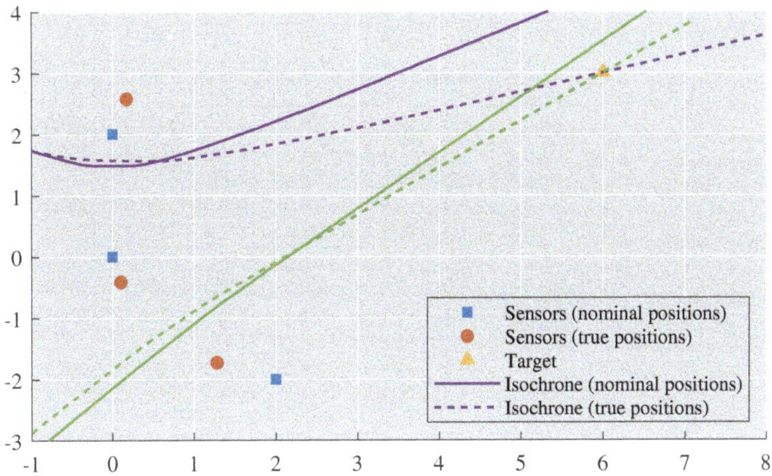

Figure 6.2 TDOA case with position uncertainty, from Example 6.2. Squares indicate the reported (nominal) positions of the TDOA sensors, while circles indicate the true (random) positions. The solid lines plot isochrones from the received data when processed with the nominal positions, while the dashed lines report the isochrones that would result from using the true positions.

We will use $\boldsymbol{\vartheta}$ in the remainder of this chapter as shorthand for the full set of unknowns. We define the Jacobian matrix of the sensor measurements $\mathbf{z}(\boldsymbol{\vartheta})$

$$\mathbf{J}_{\mathbf{z}}(\boldsymbol{\vartheta}) = \begin{bmatrix} \nabla_{\mathbf{x}}\mathbf{z}^T(\boldsymbol{\vartheta}) \\ \nabla_{\boldsymbol{\alpha}}\mathbf{z}^T(\boldsymbol{\vartheta}) \\ \nabla_{\boldsymbol{\beta}}\mathbf{z}^T(\boldsymbol{\vartheta}) \end{bmatrix} \tag{6.15}$$

To facilitate this approach, we define the gradients in (6.15) for each sensor modality, which are summarized in Table 6.2. This is implemented in the function `jacobianUnc` of the `triang`, `tdoa`, `fdoa`, and `hybrid` packages.

Table 6.2
Measurement Function Gradients

Gradient	Definition	Function
$\nabla_{\mathbf{x}}\mathbf{z}^T(\boldsymbol{\vartheta})$	(6.16)	`triang.grad_x`
	(6.27)	`tdoa.grad_x`
	(6.33)	`fdoa.grad_x`
	(6.43)	`hybrid.grad_x`
$\nabla_{\boldsymbol{\alpha}}\mathbf{z}^T(\boldsymbol{\vartheta})$	(6.24)	`triang.grad_a`
	(6.31)	`tdoa.grad_a`
	(6.42)	`fdoa.grad_a`
	(6.44)	`hybrid.grad_a`
$\nabla_{\boldsymbol{\beta}}\mathbf{z}^T(\boldsymbol{\vartheta})$	(6.21)	`triang.grad_b`
	(6.29)	`tdoa.grad_b`
	(6.36)	`fdoa.grad_b`
	(6.45)	`hybrid.grad_b`

6.1.1.1 Jacobian of AOA with Uncertainties

Given the angle measurement functions in (6.8-6.9), we can show that the gradient of each measurement with respect to target position $\mathbf{x}$ are given:

$$\nabla_{\mathbf{x}}\mathbf{z}^T(\boldsymbol{\vartheta}) = \left[\nabla_{\mathbf{x}}\boldsymbol{\psi}^T(\boldsymbol{\vartheta}), \nabla_{\mathbf{x}}\boldsymbol{\phi}^T(\boldsymbol{\vartheta})\right] \tag{6.16}$$

$$\nabla_{\mathbf{x}}\boldsymbol{\psi}^T(\boldsymbol{\vartheta}) = \left[\nabla_{\mathbf{x}}\psi_0(\boldsymbol{\vartheta}), \dots, \nabla_{\mathbf{x}}\psi_{N_a-1}(\boldsymbol{\vartheta})\right] \tag{6.17}$$

$$\nabla_{\mathbf{x}}\boldsymbol{\phi}^T(\boldsymbol{\vartheta}) = \left[\nabla_{\mathbf{x}}\phi_0(\boldsymbol{\vartheta}), \dots, \nabla_{\mathbf{x}}\phi_{N_a-1}(\boldsymbol{\vartheta})\right] \tag{6.18}$$

$$\nabla_{\mathbf{x}}\psi_n(\boldsymbol{\vartheta}) = \frac{1}{\|\mathbf{x} - \overline{\mathbf{x}}_n\|^2}\begin{bmatrix} -(y - \overline{y}_n) \\ (x - \overline{x}_n) \\ 0 \end{bmatrix} \tag{6.19}$$

$$\nabla_{\mathbf{x}}\phi_n(\boldsymbol{\vartheta}) = \frac{1}{R_{g,n}(\mathbf{x})\,\|\mathbf{x} - \overline{\mathbf{x}}_n\|^2}\begin{bmatrix} -(x - \overline{x}_n)(z - \overline{z}_n) \\ -(y - \overline{y}_n)(z - \overline{z}_n) \\ R_{g,n}^2(\mathbf{x}) \end{bmatrix} \tag{6.20}$$

In addition to the gradient with respect to target position $\mathbf{x}$, we now also take the gradient with respect to the sensor position $\overline{\mathbf{x}}$. Here, we note that the gradient of the mth measurement (ψ_m and ϕ_m) with respect to the nth sensor position error is zero if $n \neq m$, since a given sensor's AOA measurement depends only on that sensor's position, and that it is the negative of the gradient with respect to the target's position if $n = m$. We construct the gradient of all angle measurements with respect to sensor position as a function of the gradient with respect to target position (for convenience).

$$\nabla_{\boldsymbol{\beta}}\mathbf{z}^T(\boldsymbol{\vartheta}) = \left[\nabla_{\overline{\mathbf{x}}}\boldsymbol{\psi}^T(\boldsymbol{\vartheta}), \nabla_{\overline{\mathbf{x}}}\boldsymbol{\phi}^T(\boldsymbol{\vartheta})\right] \tag{6.21}$$

$$\nabla_{\overline{\mathbf{x}}}\boldsymbol{\psi}^T = \text{diag}\left\{-\nabla_{\mathbf{x}}\psi_1(\mathbf{x}), \dots, -\nabla_{\mathbf{x}}\psi_{N_a}(\mathbf{x})\right\} \tag{6.22}$$

$$\nabla_{\overline{\mathbf{x}}}\boldsymbol{\phi}^T = \text{diag}\left\{-\nabla_{\mathbf{x}}\phi_1(\mathbf{x}), \dots, -\nabla_{\mathbf{x}}\phi_{N_a}(\mathbf{x})\right\} \tag{6.23}$$

In a similar manner, the gradient of the sensor measurements with respect to the measurement biases is 1 if the bias is from the measurement in question ($m = n$) and 0 otherwise. The result is that the Jacobian is an identity matrix with $2N_a$ rows and $2N_a$ columns (assuming that the AOA measurements are reported in azimuth and elevation for all N_a AOA sensors).

$$\nabla_{\boldsymbol{\alpha}}\mathbf{z}^T(\boldsymbol{\vartheta}) = \begin{bmatrix} \nabla_{\overline{\boldsymbol{\psi}}}\boldsymbol{\psi}^T(\boldsymbol{\vartheta}) & \mathbf{0}_{N_a} \\ \mathbf{0}_{N_a} & \nabla_{\overline{\boldsymbol{\phi}}}\boldsymbol{\phi}^T(\boldsymbol{\vartheta}) \end{bmatrix} \tag{6.24}$$

$$\nabla_{\overline{\boldsymbol{\psi}}}\boldsymbol{\psi}^T(\boldsymbol{\vartheta}) = \mathbf{I}_{N_a} \tag{6.25}$$

$$\nabla_{\overline{\boldsymbol{\phi}}}\boldsymbol{\phi}^T(\boldsymbol{\vartheta}) = \mathbf{I}_{N_a} \tag{6.26}$$

6.1.1.2 Jacobian of TDOA with Uncertainties

The Jacobian matrix for TDOA measurements $\mathbf{z} = \mathbf{r}$ with sensor uncertainties takes the same form as in Chapter 2, with the partial derivatives rewritten to explicitly reference the sensor position errors $\boldsymbol{\beta} = \overline{\mathbf{x}}$ and measurement biases $\boldsymbol{\alpha} = \overline{\mathbf{t}}$:

$$\nabla_{\mathbf{x}}\mathbf{z}^T(\boldsymbol{\vartheta}) = \left[\nabla_{\mathbf{x}}R_0(\boldsymbol{\vartheta}), \dots, \nabla_{\mathbf{x}}R_{M_t-1}(\boldsymbol{\vartheta})\right] \tag{6.27}$$

$$\nabla_{\mathbf{x}}R_m(\boldsymbol{\vartheta}) = \mathbf{u}_{a_{t,m}}(\mathbf{x}, \boldsymbol{\beta}) - \mathbf{u}_{b_{t,m}}(\mathbf{x}, \boldsymbol{\beta}) \tag{6.28}$$

The gradient with respect to sensor position $\boldsymbol{\beta} = \overline{\mathbf{x}}$ is similar to that with respect to target position, except that each sensor position error occurs in only one of the two compared ranges, written:

$$\nabla_{\boldsymbol{\beta}}\mathbf{z}^T(\boldsymbol{\vartheta}) = \left[\nabla_{\overline{\mathbf{x}}}R_0(\boldsymbol{\vartheta}), \dots, \nabla_{\overline{\mathbf{x}}}R_{M_t-1}(\boldsymbol{\vartheta})\right] \tag{6.29}$$

$$\nabla_{\overline{\mathbf{x}}_n}R_m(\boldsymbol{\vartheta}) = \begin{cases} -\mathbf{u}_n(\mathbf{x}, \boldsymbol{\beta}) & n = a_{t,m} \\ \mathbf{u}_n(\mathbf{x}, \boldsymbol{\beta}) & n = b_{t,m} \\ 0 & \text{else} \end{cases} \tag{6.30}$$

Recall that $\mathbf{u}_n(\mathbf{x}, \boldsymbol{\beta})$ is the unit vector from the nth sensor to the target position $\mathbf{x}$. The shape of (6.29) is complex, but can be understood as a matrix with M_t columns and N_dN_t rows, made up of a series of $N_d \times 1$ submatrices, two per column. The

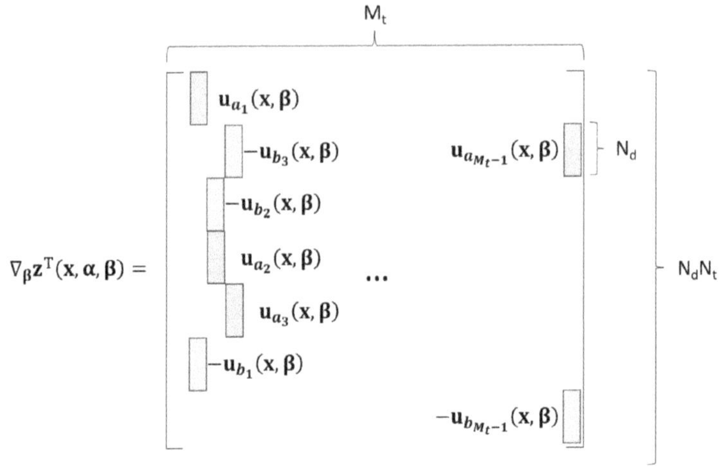

Figure 6.3 Illustration of the structure of the gradient of sensor measurements with respect to sensor position uncertainty.

rows correspond to the x/y or $x/y/z$ positions of each of the N_t TDOA sensors, while the columns correspond to a measurement. If the nth sensor is used for the mth measurement, then the cross (of that set of rows and the mth column) will be occupied either with $\mathbf{u}_n(\mathbf{x}, \boldsymbol{\beta})$ if it is a test sensor for that measurement or $-\mathbf{u}_n(\mathbf{x}, \boldsymbol{\beta})$ if it is a reference sensor for that measurement. All other elements of the matrix are 0. This structure is shown graphically in Figure 6.3.

In a similar manner, the gradient of the sensor measurements with respect to the measurement biases $\boldsymbol{\alpha} = \bar{\mathbf{r}}$ is 1 if the nth sensor is the test sensor for the mth measurement $(n = a_{t,m})$, -1 if it is the reference sensor $(n = b_{t,m})$, and 0 otherwise. The result is that the gradient is the TDOA mapping matrix $\mathbf{A}_t$ from (6.3).

$$\nabla_{\boldsymbol{\alpha}} \mathbf{z}^T(\boldsymbol{\vartheta}) = \left[\nabla_{\bar{\mathbf{r}}} R_0(\boldsymbol{\vartheta}), \dots, \nabla_{\bar{\mathbf{r}}} R_{M_t-1}(\boldsymbol{\vartheta})\right] \tag{6.31}$$

$$\nabla_{\bar{\mathbf{t}}_n} R_m(\boldsymbol{\vartheta}) = \mathbf{A}_t = \begin{cases} 1 & n = a_{t,m} \\ -1 & n = b_{t,m} \\ 0 & \text{else} \end{cases} \tag{6.32}$$

6.1.1.3 Jacobian of FDOA with Uncertainties

The Jacobian matrix for FDOA measurements $\mathbf{z} = \dot{\mathbf{r}}$ with uncertainties takes the same form as in Chapter 2, with the partial derivatives rewritten to explicitly reference the sensor position and velocity errors $\boldsymbol{\beta} = \left[\overline{\mathbf{x}}^T, \overline{\mathbf{v}}^T\right]^T$, and measurement biases $\boldsymbol{\alpha} = \dot{\overline{\mathbf{r}}}$.

$$\nabla_{\mathbf{x}}\mathbf{z}^T(\boldsymbol{\vartheta}) = \left[\nabla_{\mathbf{x}}\dot{R}_0(\boldsymbol{\vartheta}), \dots, \nabla_{\mathbf{x}}\dot{R}_{M_f-1}(\boldsymbol{\vartheta})\right] \tag{6.33}$$

$$\begin{aligned}\nabla_{\mathbf{x}}\dot{R}_m(\boldsymbol{\vartheta}) &= \mathbf{P}^{\perp}_{a_{f,m}}(\mathbf{x}, \boldsymbol{\beta})\frac{\overline{\mathbf{v}}_{a_{f,m}}}{\left\|\mathbf{x} - \overline{\mathbf{x}}_{a_{f,m}}\right\|} \\ &\quad - \mathbf{P}^{\perp}_{b_{f,m}}(\mathbf{x}, \boldsymbol{\beta})\frac{\overline{\mathbf{v}}_{b_{f,m}}}{\left\|\mathbf{x} - \overline{\mathbf{x}}_{b_{f,m}}\right\|}\end{aligned} \tag{6.34}$$

$$\mathbf{P}^{\perp}_n(\mathbf{x}, \boldsymbol{\beta}) = \mathbf{I} - \mathbf{u}_n(\mathbf{x}, \boldsymbol{\beta})\mathbf{u}_n^T(\mathbf{x}, \boldsymbol{\beta}) \tag{6.35}$$

The gradient with respect to sensor position and velocity errors is broken into two components.

$$\nabla_{\boldsymbol{\beta}}\mathbf{z}^T(\boldsymbol{\vartheta}) = \begin{bmatrix} \nabla_{\overline{\mathbf{x}}}\dot{\mathbf{r}}^T(\boldsymbol{\vartheta}) \\ \nabla_{\overline{\mathbf{v}}}\dot{\mathbf{r}}^T(\boldsymbol{\vartheta}) \end{bmatrix} \tag{6.36}$$

Just as before, the gradient with respect to sensor position errors $\overline{\mathbf{x}}$ is defined by a piecewise function

$$\nabla_{\overline{\mathbf{x}}}\dot{\mathbf{r}}^T(\boldsymbol{\vartheta}) = \left[\nabla_{\overline{\mathbf{x}}}\dot{R}_0(\boldsymbol{\vartheta}), \dots, \nabla_{\overline{\mathbf{x}}}\dot{R}_{M_f-1}(\boldsymbol{\vartheta})\right] \tag{6.37}$$

$$\nabla_{\overline{\mathbf{x}}_n}\dot{R}_m(\boldsymbol{\vartheta}) = \begin{cases} -P^{\perp}_n(\mathbf{x}, \boldsymbol{\beta})\frac{\mathbf{v}_n}{\|\mathbf{x}-\mathbf{x}_n\|} & n = a_{f,m} \\ P^{\perp}_n(\mathbf{x}, \boldsymbol{\beta})\frac{\mathbf{v}_n}{\|\mathbf{x}-\mathbf{x}_n\|} & n = b_{f,m} \\ 0 & \text{else} \end{cases} \tag{6.38}$$

In each case, the gradient is the negative of one of the two components of (6.34), depending on which sensor position is being referenced. The structure of (6.37) follows the same general structure as shown in Figure 6.3, with N_dN_f rows for the x/y or $x/y/z$ components of each of the N_f FDOA sensors, and M_f columns, one for each of the FDOA measurements.

We compute the gradient with respect to sensor velocity errors $\overline{\mathbf{v}}$ in a similar manner

$$\nabla_{\overline{\mathbf{v}}}\dot{\mathbf{r}}^T(\boldsymbol{\vartheta}) = \left[\nabla_{\overline{\mathbf{v}}}\dot{R}_0(\boldsymbol{\vartheta}), \dots, \nabla_{\overline{\mathbf{v}}}\dot{R}_{M_f-1}(\boldsymbol{\vartheta})\right] \tag{6.39}$$

$$\nabla_{\overline{\mathbf{v}}_n}\dot{R}_m(\boldsymbol{\vartheta}) = \begin{cases} -\mathbf{u}_n(\mathbf{x}, \boldsymbol{\beta}) & n = a_{f,m} \\ \mathbf{u}_n(\mathbf{x}, \boldsymbol{\beta}) & n = b_{f,m} \\ 0 & \text{else} \end{cases} \tag{6.40}$$

Just as with TDOA, the gradient with respect to measurement errors is the mapping matrix, this time $\mathbf{A}_f$ from (6.4).

$$\nabla_{\boldsymbol{\alpha}}\dot{\mathbf{r}}^T(\boldsymbol{\vartheta}) = \left[\nabla_{\bar{\dot{\mathbf{r}}}}\dot{R}_0(\boldsymbol{\vartheta}), \dots, \nabla_{\bar{\dot{\mathbf{r}}}}\dot{R}_{M_f-1}(\boldsymbol{\vartheta})\right] \tag{6.41}$$

$$\left[\nabla_{\bar{\dot{\mathbf{r}}}}R_m(\boldsymbol{\vartheta})\right]_n = \mathbf{A}_f = \begin{cases} -1 & n = a_{f,m} \\ 1 & n = b_{f,m} \\ 0 & \text{else} \end{cases} \tag{6.42}$$

6.1.1.4 Combined Jacobian

For a hybrid case, with many types of measurements, we stack the gradients defined in previous sections. The joint measurement vector is defined in (6.7), and the parameter vector to be estimated is $\boldsymbol{\vartheta}$.

The gradient with respect to $\mathbf{x}$ for all of the sensor measurements is collected

$$\nabla_{\mathbf{x}}\mathbf{z}^T(\boldsymbol{\vartheta}) = \left[\nabla_{\mathbf{x}}\boldsymbol{\psi}^T(\boldsymbol{\vartheta}), \nabla_{\mathbf{x}}\boldsymbol{\phi}^T(\boldsymbol{\vartheta}), \nabla_{\mathbf{x}}\mathbf{r}^T(\boldsymbol{\vartheta}), \nabla_{\mathbf{x}}\dot{\mathbf{r}}^T(\boldsymbol{\vartheta})\right], \tag{6.43}$$

given the gradients defined in Table 6.2. For the gradient with respect to sensor measurement biases, we first expand the bias term $\boldsymbol{\alpha} = \left[\bar{\boldsymbol{\psi}}^T, \bar{\boldsymbol{\phi}}^T, \bar{\mathbf{r}}^T, \bar{\dot{\mathbf{r}}}^T\right]^T$. The result can be shown to be:

$$\nabla_{\boldsymbol{\alpha}}\mathbf{z}^T(\boldsymbol{\vartheta}) = \mathbf{A} \tag{6.44}$$

where $\mathbf{A}$ is defined in (6.2). For the gradient with respect to sensor positions and velocities $\boldsymbol{\beta} = \left[\bar{\mathbf{x}}^T, \bar{\mathbf{v}}^T\right]^T$, we assemble the components from Table 6.2.

$$\nabla_{\boldsymbol{\beta}}\mathbf{z}^T(\boldsymbol{\vartheta}) = \begin{bmatrix} \nabla_{\bar{\mathbf{x}}}\boldsymbol{\psi}^T(\boldsymbol{\vartheta}) & \nabla_{\bar{\mathbf{x}}}\boldsymbol{\phi}^T(\boldsymbol{\vartheta}) & \nabla_{\bar{\mathbf{x}}}\mathbf{r}^T(\boldsymbol{\vartheta}) & \nabla_{\bar{\mathbf{x}}}\dot{\mathbf{r}}^T(\boldsymbol{\vartheta}) \\ \mathbf{0}_{N_dN_f,N_a} & \mathbf{0}_{N_dN_f,N_a} & \mathbf{0}_{N_dN_f,M_t} & \nabla_{\bar{\mathbf{v}}}\dot{\mathbf{r}}^T(\boldsymbol{\vartheta}) \end{bmatrix} \tag{6.45}$$

The full Jacobian matrix $J_{\mathbf{z}}(\boldsymbol{\vartheta})$ is assembled:

$$\mathbf{J}_{\mathbf{z}}(\boldsymbol{\vartheta}) = \begin{bmatrix} \nabla_{\mathbf{x}}\mathbf{z}^T(\boldsymbol{\vartheta}) \\ \nabla_{\boldsymbol{\alpha}}\mathbf{z}^T(\boldsymbol{\vartheta}) \\ \nabla_{\boldsymbol{\beta}}\mathbf{z}^T(\boldsymbol{\vartheta}) \end{bmatrix} \tag{6.46}$$

6.1.2 Sensor Position Covariance

For simplicity, we assume that the sensors can be divided into disjoint sets, with N_a AOA sensors, N_t TDOA sensors, and N_f FDOA sensors. Thus, the full set $\boldsymbol{\beta}$ is written:

$$\boldsymbol{\beta} = \begin{bmatrix} \mathbf{x}_a \\ \mathbf{x}_t \\ \mathbf{x}_f \\ \mathbf{v}_f \end{bmatrix} \tag{6.47}$$

where $\mathbf{x}_a$ is the $N_d N_a \times 1$ vector of AOA sensor positions, $\mathbf{x}_t$ is the TDOA sensor positions, and $\mathbf{x}_f$ is the vector of FDOA sensor positions, with velocities $\mathbf{v}_f$. This simplifies the notation within this chapter, and works well with the existing MATLAB® codebase from [1], which specifies sensor positions separately for AOA, TDOA, and FDOA.

Previously, colocated sensors (such as a joint TDOA/FDOA sensor or joint AOA/TDOA sensor) did not pose a problem, as the coordinates for that sensor appear in more than one sets of positions. Unfortunately, if we are estimating sensor positions, it is critical that we understand the relationships between each of these sensors. To handle this case, we will reflect the dependence of these two sensors via their covariance matrix.

Recall that we assume $\boldsymbol{\beta}$ is distributed as a Gaussian random vector:

$$\boldsymbol{\beta} \sim \mathcal{N}\left(\boldsymbol{\beta}^0, \mathbf{C}_{\boldsymbol{\beta}}\right) \tag{6.48}$$

where $\boldsymbol{\beta}^0$ is the reported sensor position (previously assumed to be exact), and $\mathbf{C}_{\boldsymbol{\beta}}$ is the variance of $\boldsymbol{\beta}$. If two sensors are colocated on the same platform, then that will be reflected in the structure of $\mathbf{C}_{\boldsymbol{\beta}}$.

First, we consider the case where every sensor is independent, on a separate platform. In this case, $\mathbf{C}_{\boldsymbol{\beta}}$ is a block diagonal matrix

$$\mathbf{C}_{\boldsymbol{\beta}} = \operatorname{diag}\left\{\mathbf{C}_0, \mathbf{C}_1, \dots, \mathbf{C}_{N_a+N_t+2*N_f}\right\} \tag{6.49}$$

where $\mathbf{C}_n$ is the $N_d \times N_d$ covariance matrix of the x/y or $x/y/z$ components of the nth sensor.[3] If, however, there is a sensor that is present in more than one set, then the cross terms must reflect that correlation. In Figure 6.4, we demonstrate the structure of a set of sensors with one repeated sensor between AOA and TDOA.

To generate $\mathbf{C}_{\boldsymbol{\beta}}$, using the structure in Figure 6.4, we have implemented the utility `utils.parsePosVelCovar`. This utility takes in a single vector of sensor positions, and optional velocities (if FDOA is used), as well as a set of indices for AOA, TDOA, and FDOA. These are used to generate the components of $\boldsymbol{\beta}^0$. Then, if a covariance matrix is defined for each sensor (currently as a 3-D matrix $N_d \times N_d \times N$ for N unique sensors), the indices and covariances are used to generate $\mathbf{C}_{\boldsymbol{\beta}}$. Example usage is:

```
[x_aoa, x_tdoa, x_fdoa, v_fdoa, C_beta] = ...
        parsePosVelCovar(x_sensor, v_sensor, aoa_idx, tdoa_idx,...
                        fdoa_idx, C_pos, C_vel);
```

Example 6.3 Joint AOA/TDOA with Bias and Position Uncertainty

Consider the AOA system from Example 6.1 and the TDOA system from Example 6.2. Assume that the two AOA sensors that are colocated with two of the TDOA sensors are

3. This form assumes that for the N_f FDOA sensors, their positions and velocities are independent of each other.

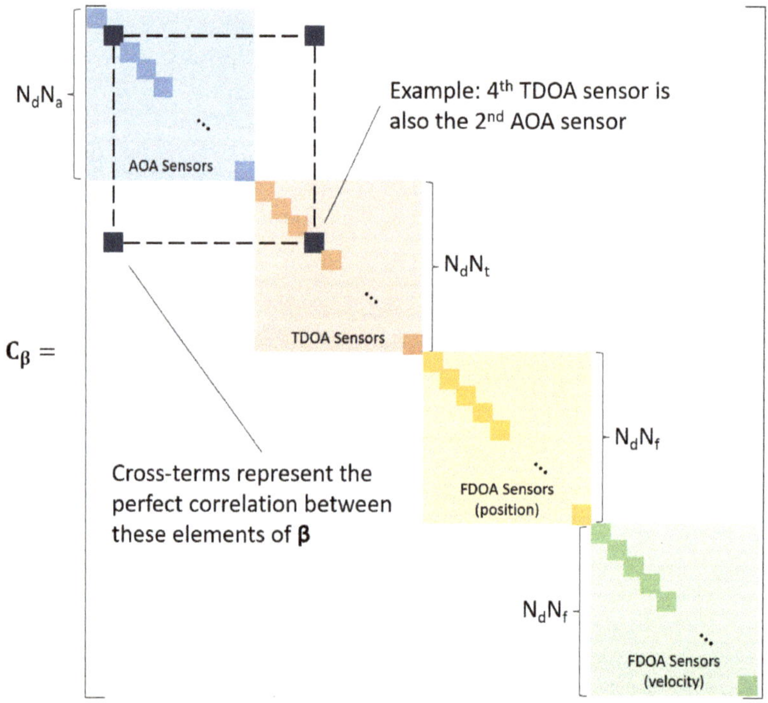

Figure 6.4 Illustration of the structure of $\mathbf{C}_\beta$ with a sensor that generates both AOA and TDOA measurements.

on the same platform (that is, their positions are fully correlated) and that the remaining AOA platform has independent sensor positions errors distributed similar to the TDOA platforms. In other words, there are two joint TDOA/AOA sensors, one AOA-only sensor, and two TDOA-only sensors. What is the joint sensor covariance matrix $\mathbf{C}_\beta$?

Generate a set of AOA and TDOA measurements that include sensor position uncertainty, as well as the AOA bias from Example 6.1 (but no TDOA bias) from a single transmitter at position (0, 5) km. Compare to the ideal measurements for that transmitter.

Solution: To begin, let's identify the paired AOA/TDOA sensors. We'll do this with a call to `utils.rng`. The results are fed to `ind2sub` to generate a pair of index vectors into the AOA and TDOA sensors.

```
dist = utils.rng(x_aoa,x_tdoa);
idx_pairs = find(dist(:)==0);
[idx_aoa,idx_tdoa] = ind2sub(size(dist),idx_pairs);
```

The ith element of `idx_aoa` and `idx_tdoa` are indices to the matching sensors; those at the same position. In our case, the second AOA sensor aligns with the second TDOA sensor, and the third AOA sensor with the third TDOA sensor.

```
> idx_aoa = [2, 3]
> idx_tdoa = [2, 3]
```

Next, we build the covariance matrix for position uncertainty to match the structure in Figure 6.4. This is a 3-step process. First, we'll define a covariance matrix `C_pos_1d` that reflects the structure of Figure 6.4, but with unit variance, and only one dimension. Then, we'll define a smaller covariance matrix that reflects the covariance of position uncertainty for the spatial coordinates of a single sensor. Then, we'll combine the two with a *kronecker product*.[4]

```
C_pos_1d = eye(n_tdoa+n_aoa); % diagonal entries of Figure 6.4
for ii = 1:numel(idx_pairs)
    this_idx_aoa = idx_aoa(ii);
    this_idx_tdoa = idx_tdoa(ii);
    C_pos_1d(this_idx_aoa,n_aoa+this_idx_tdoa) = 1; % build off-diagonal
    C_pos_1d(n_aoa+this_idx_tdoa,this_idx_aoa) = 1; % entries
end

C_beta_single = .1*eye(num_dims); % single sensor covariance
C_beta = kron(C_pos_1d,C_beta_single);
```

The initial covariance matrix is an identity matrix with additional ones aligning the second AOA with the second TDOA sensor (colored in red) and the third AOA with

4. The kronecker product `kron(a,b)` takes the matrix `b` and multiplies it against each element of `a`, resulting in a larger matrix.

the third TDOA sensor (colored in blue):

$$C\_pos\_1d = \begin{bmatrix} 1 & 0 & 0 & 0 & 0 & 0 \\ 0 & 1 & 0 & 0 & 1 & 0 \\ 0 & 0 & 1 & 0 & 0 & 1 \\ 0 & 0 & 0 & 1 & 0 & 0 \\ 0 & 1 & 0 & 0 & 1 & 0 \\ 0 & 0 & 1 & 0 & 0 & 1 \end{bmatrix} \tag{6.50}$$

The full covariance matrix is:

$$\mathbf{C}_\beta = \begin{bmatrix} .1 & 0 & 0 & 0 & 0 & 0 & 0 & 0 & 0 & 0 & 0 & 0 \\ 0 & .1 & 0 & 0 & 0 & 0 & 0 & 0 & 0 & 0 & 0 & 0 \\ 0 & 0 & .1 & 0 & 0 & 0 & 0 & 0 & .1 & 0 & 0 & 0 \\ 0 & 0 & 0 & .1 & 0 & 0 & 0 & 0 & 0 & .1 & 0 & 0 \\ 0 & 0 & 0 & 0 & .1 & 0 & 0 & 0 & 0 & 0 & .1 & 0 \\ 0 & 0 & 0 & 0 & 0 & .1 & 0 & 0 & 0 & 0 & 0 & .1 \\ 0 & 0 & 0 & 0 & 0 & 0 & .1 & 0 & 0 & 0 & 0 & 0 \\ 0 & 0 & 0 & 0 & 0 & 0 & 0 & .1 & 0 & 0 & 0 & 0 \\ 0 & 0 & .1 & 0 & 0 & 0 & 0 & 0 & .1 & 0 & 0 & 0 \\ 0 & 0 & 0 & .1 & 0 & 0 & 0 & 0 & 0 & .1 & 0 & 0 \\ 0 & 0 & 0 & 0 & .1 & 0 & 0 & 0 & 0 & 0 & .1 & 0 \\ 0 & 0 & 0 & 0 & 0 & .1 & 0 & 0 & 0 & 0 & 0 & .1 \end{bmatrix} \tag{6.51}$$

The next step is to generate measurements. We've discussed previously how to use the Cholesky decomposition to generate correlated Gaussian random variables, but the covariance matrix we have here is singular; not all of its eigenvalues are positive. In this case, the Cholesky decomposition fails. An alternative approach is to the use the *singular value decomposition*, and to use the left singular vectors along with the square roots of the eigenvalues.[5.]

```
[U,S,~] = svd(C_beta);
epsilon = reshape(U*sqrt(S)*randn(num_dims*(n_aoa+n_tdoa),1),num_dims,n_aoa+
    n_tdoa);
```

This creates a $2 \times$ `n_aoa` $+$ `n_tdoa` matrix of perturbations. Finally, we separate the perturbations into those that apply to the AOA sensors and those for the TDOA sensors.

```
beta_aoa = x_aoa + epsilon(:,1:n_aoa);
beta_tdoa = x_tdoa + epsilon(:,n_aoa + (1:n_tdoa));
```

To verify that the generated positions are appropriately correlated, we'll recheck the distance between sensors.

5. To verify this, one should change the definition of `epsilon` to generate a large number of columns (at least one million), then compute the sample correlation matrix `C_est = cov(epsilon');`, and compare the result to `C_beta`.

```
dist_perturbed = utils.rng(beta_aoa,beta_tdoa);
assert(all(dist_perturbed(idx_pairs)<1e-6),'Error generating correlated
    sensor perturbations.');
```

The lack of an error message in our example confirms they were generated correctly. In other words, the position for the second AOA and second TDOA sensors is within 1 μm of each other, and the third AOA and third TDOA sensors are similarly close.

The next step is to generate measurements and compare them. We can use the `hybrid.measurement` function to generate three sets: (1) ideal measurements, (2) position uncertainty but not bias, and (3) both position uncertainty and bias.

```
x_tgt = [6; 3];

alpha_aoa = [5, 10, -5]*pi/180; % AOA bias
% truth
zeta = hybrid.measurement(x_aoa,x_tdoa,[],[], x_tgt, n_tdoa);
% with position uncertainty
zeta_unc = hybrid.measurement(beta_aoa,beta_tdoa,[],[],x_tgt,n_tdoa);
% with position uncertainty and measurement bias
zeta_unc_bias = hybrid.measurement(beta_aoa,beta_tdoa,[],[], x_tgt, n_tdoa
    ,[],[],alpha_aoa);
```

The result is:

```
Measurements from ideal sensors (AOA, AOA, AOA, RDOA, RDOA):
[0.24, 0.79, 0.46, -0.63, -1.05]
With pos unc:
[0.30, 0.75, 0.53, -0.96, -0.83]
With pos unc and bias:
[0.39, 0.84, 0.62, -0.96, -0.83]
```

The first thing to notice is that all of the measurements changed when the position uncertainty is taken into effect, while only the AOA measurements are changed when the position bias is added in. Figure 6.5 plots the three sets of measurements, and the lines of bearing and isochrones that result from processing the noisy and biased measurements at the (incorrect) nominal sensor positions.

□

6.2 APPROACH

It may be tempting, even for small biases, to simply ignore them and solve the geolocation problem as if they were not present. In some cases, this may be acceptable, such as when all sensors are shifted some small distance in the same direction, as shown in

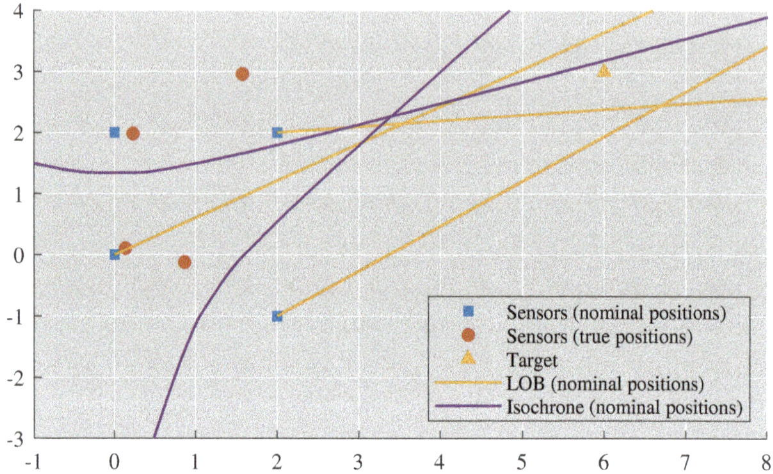

Figure 6.5 Joint AOA/TDOA case with position uncertainty, from Example 6.3. Squares indicate the reported (nominal) positions of the AOA/TDOA sensors, while circles indicate the true (random) positions. The solid lines plot the lines of bearing and isochrones from the received data when processed with the nominal positions, while the dashed lines report the lines of bearing and isochrones that would result from using the true positions.

Figure 6.6(a).[6] In this case, the error in geolocation is on the same order as the errors in sensor position. If the errors are not aligned, however, we can experience geolocation error that is much larger than the sensor position errors, such as is shown in Figure 6.6(b). It is, therefore, critical to understand and properly model systematic biases.

We go into the relationship between sensor position errors and geolocation errors in the video linked in Figure 6.7.

6.2.1 Required Observations

The first step in selecting a solution is to examine whether the problem is over- or under-determined. For example, in [11], the authors consider a 1-D AOA problem with N sources and M sensors (with perfect sensor position but unknown measurement bias) and find that $MN > 2N + M$ is required for observability of the solution space. This is feasible for modestly large problems (with $M = 3$ sensors, then $N \geq 3$ sources must be observed). However, as the number of unknowns grows, such as the sensor position, it may be difficult to guarantee observability.

In general, we can construct the matrix $\mathbf{J}_\mathbf{z}(\boldsymbol{\vartheta})$ in (6.46), remove any extraneous columns (e.g., the elevation angle columns if AOA sensors are reported azimuth alone,

6. See Ma and Ho [10] for an analysis of limits on sensor position error under which they can be safely ignored.

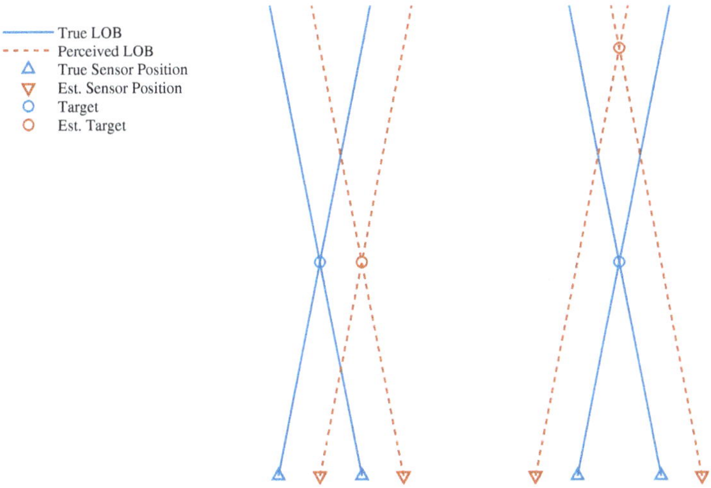

(a) (b)

Figure 6.6 Illustration of the nonlinear impact of position errors on geolocation errors. (a) Uniform sensor position errors, and (b) nonuniform sensor position errors.

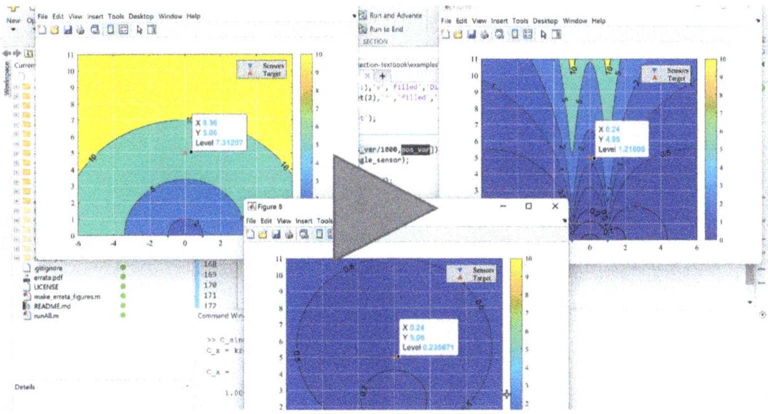

Figure 6.7 Video discussing the relationship between sensor bias errors and geolocation error.

or columns from sensors that provide only one measurement type instead of all three), and remove any extraneous rows (e.g., any terms that are not relevant, such as frequency offset in an AOA/TDOA problem, or any terms that are known to sufficient precisions, such as sensor position in a GPS-enabled scenario). The final Jacobian must have more columns than rows for the problem to be fully *observable*, corresponding to a critically or over-determined problem, as opposed to *unobservable* or under-determined, where there are more unknowns than observations.[7]

If observability is met (that is, the product of the number of measurements and sources exceeds the number of unknowns), then traditional geolocation methods are likely sufficient; we will discuss their use in Sections 6.2.2 and 6.2.3.

If observability is not met, as will often be the case when there are sensor uncertainties, then additional observations are needed. Two popular methods to achieve this are: (1) repeated observations and tracking of targets over time, and (2) the use of calibration emitters (artificial sources, potentially with known positions). We'll discuss the former when we introduce the concept of geolocation with multiple snapshots in Chapter 7, and the latter will be discussed in Section 6.3.

6.2.2 Maximum Likelihood

In this section, we discuss application of traditional geolocation algorithms to the problem at hand: geolocation with sensor parameter uncertainties. Using this formulation, one could naively solve this problem using the same maximum likelihood approach as we have used for simpler geolocation problems. The principal challenge with such an approach is that the 2-D or 3-D search grid is now replaced with a much higher dimensional search grid across $\boldsymbol{\vartheta}$. Performing a brute-force ML optimization over a grid space with increasing dimensionality like this will quickly become computationally infeasible, particularly if the search space for biases, sensor position errors, and target position is large or must be densely sampled.

Nevertheless, this can be attempted. To apply maximum likelihood, we simply replace the likelihood function with one built from (6.6), and the variable `x` in the function handle is replaced with a larger parameter vector `theta` that contains the target position, sensor positions, and measurement biases.

```
ell = @(theta) hybrid.loglikelihoodUnc(...
               x_aoa, x_tdoa, x_fdoa, v_fdoa, ... % reported positions
               zeta, C, C_beta, theta, ...
               tdoa_ref_idx, fdoa_ref_idx);
```

The variable `theta` is the full vector of unknowns $\boldsymbol{\vartheta}$.

7. Strictly speaking, it must have more *linearly independent* columns than rows [12]. This independence requirement accounts for degenerate geometries, or duplicate measurements that provide no independent information, such as when one the target and two AOA sensors lie along a line.

The search grid center th_center is then built by stacking together the provided initial estimate of target position x_init with the provided sensor positions and an initial assumption of zero bias for all measurements.

```
th_ctr = [x_ctr;
          zeros(m_aoa+m_tdoa+m_fdoa,1); % alpha = 0
          x_aoa(:);
          x_tdoa(:);
          x_fdoa(:);
          v_fdoa(:)];
```

The grid creation step of utils.mlSoln had to be adjusted to handle a parameter vector that has more than 3 elements. This is likely the place where a problem will happen if the search grid is too large, so we have added some code that checks to make sure the array sizes are not going to exceed MATLAB®'s limits. The final call results in an estimate of the ϑ, and so we parse the estimated target position from that vector.

```
[th_est,A,x_grid] = utils.mlSoln(ell,th_ctr,search_size,epsilon);
x_est = th_est(x_ind);
```

The search grid size is controlled by specifying the input search_size. For estimation of ϑ, it is imperative that we set an appropriate search size across the different dimensions and resolution. For example, we could define:

```
search_size = [10e3 * ones(1, n_dim), ...           % x [m]
               pi/180 * ones(1,m_aoa), ...         % alpha_aoa [rad]
               10 * ones(1,m_tdoa), ...            % alpha_tdoa [m]
               10 * ones(1,m_fdoa), ...            % alpha_fdoa [m/s]
               25 * ones(1, n_aoa*n_dim), ...      % x_aoa [m]
               25 * ones(1, n_tdoa*n_dim), ...     % x_tdoa [m]
               25 * ones(1, n_fdoa*n_dim), ...     % x_fdoa [m]
               10 * ones(1, n_fdoa*n_dim)];        % v_fdoa [m/s]
```

in order to search for the target within a 10 km × 10 km × 10 km region, with up to 1 degree of bias in AOA measurements, 10 m of range bias in TDOA measurements, 10 m/s of range-rate bias in FDOA measurements, 25 m of sensor position error (in any dimension) for any of the AOA, TDOA, or FDOA sensors, and 10 m/s of sensor velocity error. Similarly, the epsilon variable defines how densely the grid will be sampled. Within the mlSolnUnc function, both search_size and epsilon can either be defined for the full ϑ vector (as illustrated here), or for just the target dimensions, in which case the default values will match the code snippet above for search_size, and the resolution epsilon will be sized for a default of 11 sample points in each of the $\boldsymbol{\alpha}$ and $\boldsymbol{\beta}$ dimensions.

This will result in an unreasonably large array if ϑ has more than a handful of dimensions. For example, if the maximum array size in MATLAB® is 8 GB, then a double precision array can have no more than 10^9 elements. If the target grid is 101 ×

101, then that is 10,201 possible target locations. Even a modest 3 sensor AOA case, in 2-D, results in a 3-element vector $\boldsymbol{\alpha}$ and a 6-element vector $\boldsymbol{\beta}$. If there are 11 grid points for each element of $\boldsymbol{\alpha}$ and $\boldsymbol{\beta}$, the resulting grid will have $2.4 * 10^{13}$ elements; far larger than the maximum allowable array size. Use of ML search is only advisable in reduced uncertainty scenarios, such as when only one sensor position needs to be estimated, or when positions are known well but biases must be estimated.

Example 6.4 Likelihood with TDOA Bias

Consider a 4-sensor TDOA system, with sensors at the positions (2, 2), (0, 2), (4, 0), and (0, 0) km, with perfect position accuracy. Assume that the measurements have bias $\boldsymbol{\alpha} = [10, 30, -20, 60]\ m$, and the target is located at (6, 3) km.

Generate a set of noisy measurements with $\sigma_t^2 = (100ns)^2$ for each sensor, using the final sensor as the common reference. Compute the log likelihood for a grid of target points between (0, 0) and (10, 10) km, sampled at 0.1 km in each dimension. Plot the log likelihood with and without measurement bias. Comment on the difference in the likelihood plots.

Use the ML solver to solve for $\mathbf{x}$ and $\boldsymbol{\alpha}$, using the same grid of target position, and search for biases over the interval $\alpha_m \in [-80, 80]$ m, sampled in 10 m increments.[8] To compensate for the larger search space, use a 500 m grid spacing for the target position.

Solution: First, we generate target and sensor positions (the code is omitted, for brevity), then we define the measurement bias and generate noise-free measurements both with and without bias.

```
alpha_tdoa = [10, 30, -20, 60]'; % TOA bias
% with measurement bias
z = tdoa.measurement(x_tdoa,x_tgt, n_tdoa, alpha_tdoa);
% truth
z_true = tdoa.measurement(x_tdoa,x_tgt, n_tdoa);
```

Generating noise for the two sets of measurements is straightforward, using the Cholesky decomposition of the resampled RDOA covariance matrix.

```
err_toa = 100e-9;
C_toa = err_toa^2*eye(n_tdoa);
C_roa = utils.constants.c^2*C_toa;
C_rdoa = utils.resampleCovMtx(C_roa,n_tdoa);
L= chol(C_rdoa,'lower');

noise = L*randn(size(U,2),1);
zeta = z + noise;
zeta_true = z_true + noise;
```

8. This corresponds to a timing bias interval of $\approx [-267, 267]$ ns with a resolution of ≈ 33 ns.

Next, we define the grid of target positions and compute the log likelihood both without and with the bias terms, respectively, in Figure 6.8. In the former, Figure 6.8(a), the bias-free likelihood plot has a peak that is well aligned with the true target position, although somewhat offset by the noise included in the measurement. Figure 6.8(b) plots the likelihood function when bias terms and noise both pollute the likelihood calculation. In this case, we see that the peak is moved slightly further away from the target location. In this case, notionally, the bias term appears to have a similar impact on the position estimate as the noise does (albeit a stationary bias effect, rather than an unbiased random effect).

```
x_vec_km = 0:.1:10;
[xx,yy] = meshgrid(x_vec_km);
x_grid = [xx(:),yy(:)]'*1e3; % meters

ell = tdoa.loglikelihood(x_tdoa,zeta,C_roa,x_grid,n_tdoa);
ell_true = tdoa.loglikelihood(x_tdoa,zeta_true,C_roa,x_grid,n_tdoa);
```

We begin with a typical ML solver that is not bias-aware. In this case, for both the bias-free and bias measurements, we use the `tdoa.mlSoln` solver to estimate the target position.

```
x_ctr = [5,5]*1e3;
search_size = 10e3;
epsilon = .1e3;

x_est_true = tdoa.mlSoln(x_tdoa,zeta_true,C_roa,x_ctr,search_size,epsilon,
    n_tdoa);
x_est = tdoa.mlSoln(x_tdoa,zeta,C_roa,x_ctr,search_size,epsilon,n_tdoa);
```

Next, we define the ML solver with uncertainty. To set up this solver, we must define the search space for $\boldsymbol{\vartheta}$ by specifying the center of the search interval, the search size (maximum offset in each dimension), and the resolution. The way that `mlSolnUnc` is written, $\boldsymbol{\vartheta}$ must include both $\boldsymbol{\alpha}$ and $\boldsymbol{\beta}$ terms, regardless of whether we wish to search over both sets of uncertainties. To ignore $\boldsymbol{\beta}$, we set the search size on those elements in the vector to 0.

```
th_ctr = cat(1,x_ctr,zeros(n_tdoa,1),x_tdoa(:)); % expand parameter vector
search_size = cat(1,5e3*ones(2,1), ...        % x_tgt search
                    80*ones(n_tdoa,1), ...    % alpha search
                    zeros(numel(x_tdoa),1)); % beta search
epsilon = cat(1,500*ones(2,1), ...            % x_tgt resolution
                10*ones(n_tdoa,1), ...        % alpha resolution
                ones(numel(x_tdoa),1));       % beta resolution

C_beta = .001*eye(numel(x_tdoa)); % position covariance error
  % C_beta doesn't really matter, since we're not searching over it
```

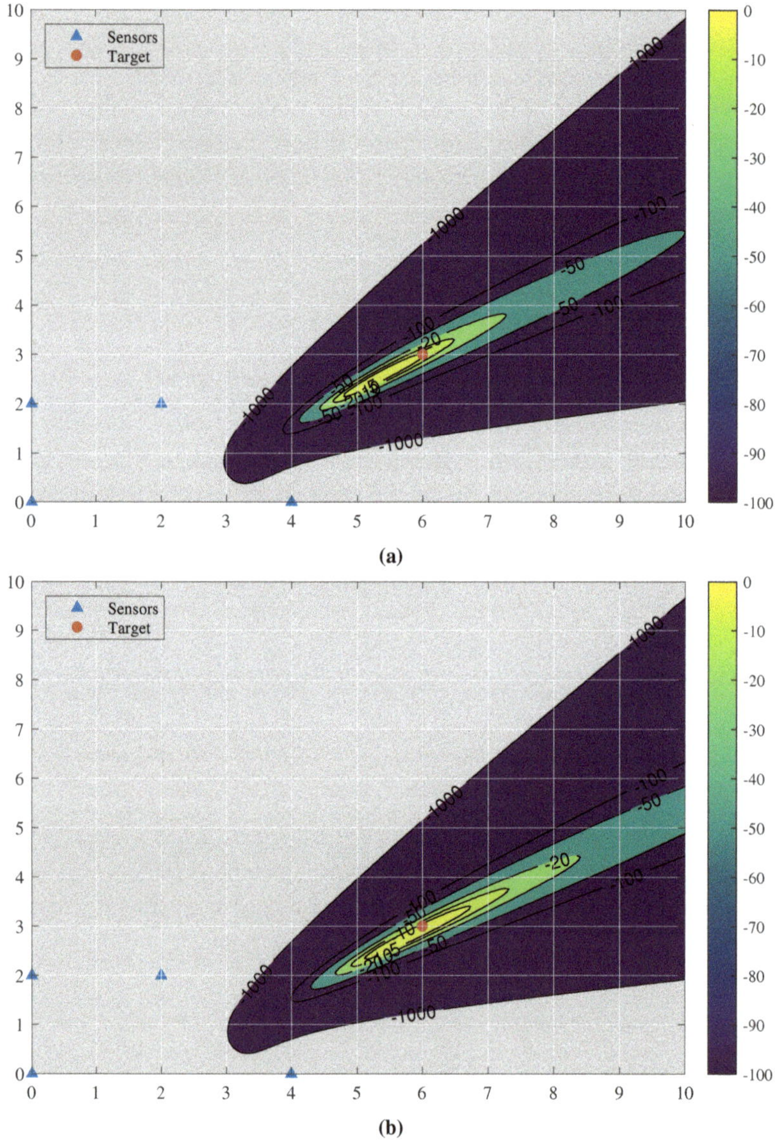

Figure 6.8 Log likelihood contours for the 4-element TDOA scenario in Example 6.4. (a) Without range bias term, and (b) with range bias terms.

```
[x_est_bias, alpha_est, ~] = tdoa.mlSolnUnc(x_tdoa, zeta, C_roa, C_beta,
    th_ctr, search_size, epsilon, n_tdoa);
```

In the original case, our 100 m spacing resulted in a 2-D grid of 10,201 positions (101 points in both x and y). In the bias estimation search, we reduced the spatial grid to only 441 positions (21 points in both x and y), but there are also 17 positions for each of the four range bias terms, leading to a total of 83,521 range bias options to search across. The total number of search points for $\boldsymbol{\vartheta}$ is, thus, just shy of 37 million. Even a modest search space like this will quickly grow unwieldy when searching over a handful of parameters.

The results are printed here, where we see that in this initial case, the range bias terms actually resulted in a *closer* geolocation estimate, as they happened to work in opposition to the noise experienced in this test. The estimator with uncertainty, however, is able to recover the true target position, beating the performance (in this isolated test case) of the bias-free estimator. These estimates, along with the log-likelihood contours originally shown in Figure 6.8 are plotted in Figure 6.9.

```
True ML Est.: (5.30, 2.50) km, error: 0.86 km
Biased ML Est.: (5.80, 2.80) km, error: 0.28 km
ML Est. w/Uncertainty: (6.00, 3.00) km, error: 0.00 km
```

The ML estimator with uncertainty also generates an estimate of the range bias terms. In this case, we see that the estimated bias terms are close, but don't line up as well as the target position estimates do. Note that we're not able to directly estimate the range bias terms, it is the range bias *difference* that can be estimated, and so the bias term for the common reference (the fourth sensor in this case) will always return zero. In this case, we'd expect range difference measurements of -50, -30, and -80 m, but instead we see range difference measurements of 10, 50, and -10.

```
True range bias: (10, 30, -20, 60) m
Estimated range bias: (10, 50, -10, 0) m
```

□

6.2.3 Iterative Solvers

The iterative approaches previously discussed (iterative least square approximation, and gradient descent) begin with the gradient of the measurements with respect to the parameter vector $\boldsymbol{\vartheta}$, which we derived in (6.46). Previously, we noted the similarity between these solutions and the gradient of the likelihood function, particularly once we appropriately scaled them by including consideration of the measurement covariance matrix $\mathbf{C}_\zeta$. This similarity means that both of those iterative algorithms function as approximations of the maximum likelihood solution, and perform very well when compared to the CRLB.

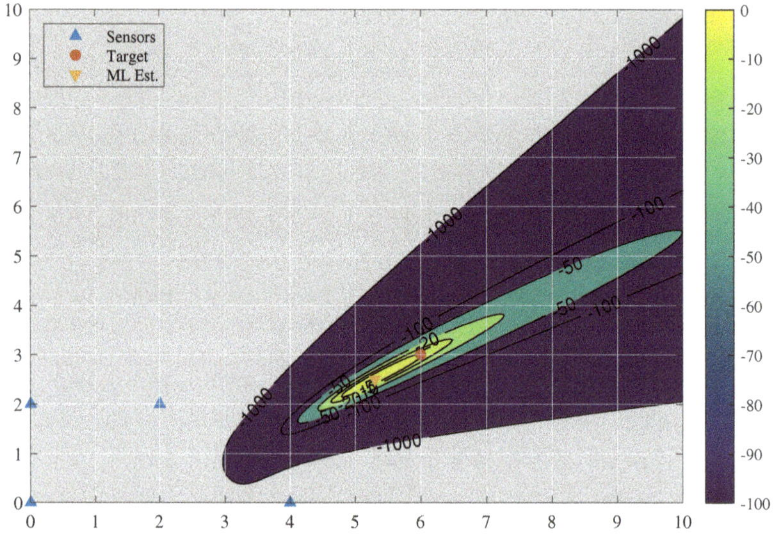

(a)

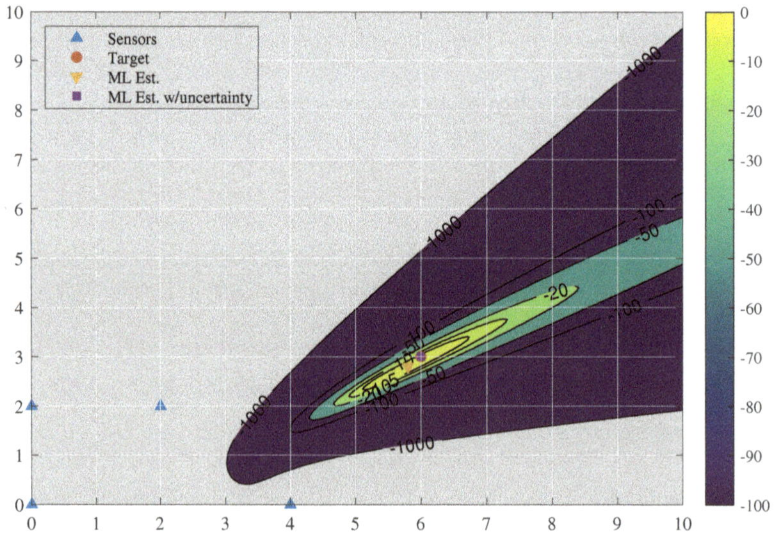

(b)

Figure 6.9 Log likelihood contours and ML estimate solutions for the scenario in Example 6.4. (a) Without range bias term, and (b) with range bias terms.

In this problem, however, the sensor position (and velocity) uncertainty manifests as a second term in the likelihood function of (6.6). That second term is not reflected in either the gradient descent or iterative least squares algorithms. This missing term means that they are no longer approximations of the maximum likelihood solution, and so we will no longer expect them to achieve the CRLB, particularly as the amount of uncertainty in sensor position increases ($\mathbf{C}_\beta$).

Nevertheless, we have implemented calling functions for both of these algorithms, as `gdSolnUnc` and `lsSolnUnc` in the `triang`, `tdoa`, `fdoa`, and `hybrid` packages, which swap out the standard Jacobian and measurement functions for those derived in this chapter.

For example, in the standard solve `hybrid.lsSoln`, the anonymous function used to compute the measurement error for an estimated target position $\hat{\mathbf{x}}$ is:

```
y = @(x) z - hybrid.measurement(x_aoa, x_tdoa, x_fdoa, v_fdoa, ...
                                x, tdoa_ref_idx, fdoa_ref_idx);
```

while the corresponding line in `hybrid.lsSolnUnc` is:

```
y = @(theta) z - hybrid.measurement(...
            reshape(theta(beta_a_ind),n_dim,n_aoa), ...    % x_aoa
            reshape(theta(beta_t_ind),n_dim,n_tdoa), ...   % x_tdoa
            reshape(theta(beta_fx_ind),n_dim,n_fdoa), ...  % x_fdoa
            reshape(theta(beta_fv_ind),n_dim,n_fdoa), ...  % v_fdoa
            reshape(theta(x_ind),n_dim,1), ...             % x
            tdoa_ref_idx, fdoa_ref_idx, ...
            reshape(theta(alpha_a_ind),n_aoa,1), ...       % alpha_aoa
            reshape(theta(alpha_t_ind),n_tdoa_msmt,1), ...% alpha_tdoa
            reshape(theta(alpha_f_ind),n_fdoa_msmt,1));    % alpha_fdoa
```

where the variable `theta` is the full vector of unknowns $\boldsymbol{\vartheta}$, just as in the modified ML solver. The initial solution `th_init` is then built the same way as `th_center` was for the ML solver (by stacking our initial guesses for target position, measurement bias, and sensor position).

The calling function is then unchanged; the standard `lsSoln` solution in the `utils` package will suffice, but with a much larger `th_init` and matching Jacobian and error functions. The result, an estimate of $\boldsymbol{\vartheta}$, is then sampled to isolate the estimated target position $\hat{\mathbf{x}}$.

```
[th,th_full] = utils.lsSoln(y,J,C_tilde,th_init,epsilon,max_num_iterations,
    force_full_calc,plot_progress);

% Grab the x coordinates
x = th(x_ind);
x_full = th_full(x_ind,:);
```

We explore an expansion of Example 6.4 with iterative solvers in the video linked in Figure 6.10.

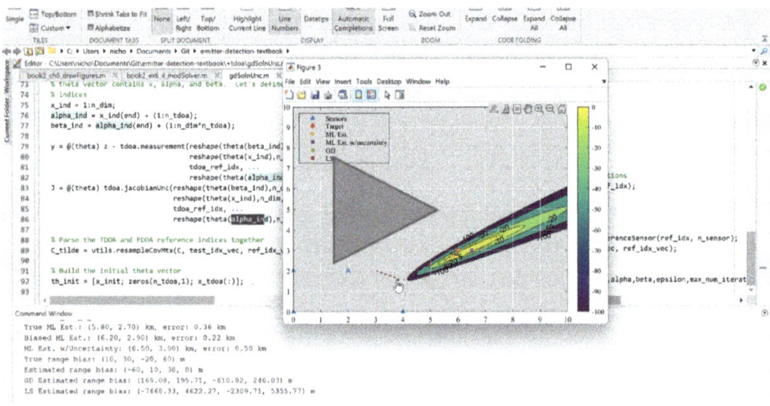

Figure 6.10 Video discussing the use of iterative solvers for the problem in Example 6.4.

6.3 CALIBRATION EMITTERS

One approach to improving performance, and overcoming the issue of extra unknowns, is to use calibration emitters. Broadly speaking, these are emitters that have a known position, and whose signals are collected and processed in the same manner as target signals, except that the known source positions are used to improve estimates of the sensor positions and measurement biases. Calibration emitters have been reported on for AOA [13], [14], TDOA [15]–[17], FDOA [18], and hybrid [19]–[21] geometries. In each case, the formulation is application-specific, and a variety of algorithms are explored. For example, some sources explicitly estimate measurement bias and sensor position error from the calibration measurements [17], [19], while others use the calibration measurements to adjust the readings from targets to compensate for those errors without explicitly estimating them [22]. We choose to use the former approach, as the latter appears to be more sensitive to variations in the source and calibration emitters (in other words, it is most accurate when the calibration emitters are near the target and, for FDOA, have similar velocities).

We will focus on a generalization of the two-stage algorithm in [17], which first uses the calibration emitter measurements to estimate source error and biases, and then uses those estimated parameters with the target measurements to estimate target location. In [17] they also include a bias estimate refinement in the second stage, but we will skip this step for simplicity. As with all of the implemented solutions, what we provide is not intended to be optimal, but rather is an attempt to provide a general solution that works across a broad set of scenarios and can be used as a benchmark against which optimized algorithms are compared.

Let $\mathbf{X}_c$ be the matrix of calibration positions and $\mathbf{z}_c(\boldsymbol{\alpha}, \boldsymbol{\beta})$ be the set of measurement functions that describe the calibration measurements

$$\mathbf{X}_c = [\mathbf{x}_{c_1}, \dots, \mathbf{x}_{c_K}] \tag{6.52}$$

$$\mathbf{z}_c(\boldsymbol{\alpha}, \boldsymbol{\beta}) = \left[\mathbf{z}^T(\mathbf{x}_{c_1}, \boldsymbol{\alpha}, \boldsymbol{\beta}), \dots, \mathbf{z}^T(\mathbf{x}_{c_1}, \boldsymbol{\alpha}, \boldsymbol{\beta})\right]^T \tag{6.53}$$

where $\mathbf{z}(\mathbf{x}_{c_k}, \boldsymbol{\alpha}, \boldsymbol{\beta})$ represents the measurement function (defined by the type of measurements taken) across all receivers for the calibration emitter located at $\mathbf{x}_{c_k}$, $k = 1, \dots, K$. Let $\boldsymbol{\zeta}_c$ be the received (noisy) calibration measurements, distributed as Gaussian random variables:

$$\boldsymbol{\zeta}_c \sim \mathcal{N}(\mathbf{z}_c(\boldsymbol{\alpha}, \boldsymbol{\beta}), \mathbf{C}_{\zeta_c}) \tag{6.54}$$

with likelihood function

$$\ell(\boldsymbol{\zeta}_c|\boldsymbol{\alpha}, \boldsymbol{\beta}) = [\boldsymbol{\zeta}_c - \mathbf{z}_c(\boldsymbol{\alpha}, \boldsymbol{\beta})]^T \mathbf{C}_{\zeta_c}^{-1} [\boldsymbol{\zeta}_c - \mathbf{z}_c(\boldsymbol{\alpha}, \boldsymbol{\beta})] \tag{6.55}$$

The gradient of the calibration measurements is

$$\nabla_{\boldsymbol{\alpha}} \mathbf{z}_c^T(\boldsymbol{\alpha}, \boldsymbol{\beta}) = \left[\nabla_{\boldsymbol{\alpha}} \mathbf{z}^T(\mathbf{x}_{c_1}, \boldsymbol{\alpha}, \boldsymbol{\beta}), \dots, \nabla_{\boldsymbol{\alpha}} \mathbf{z}^T(\mathbf{x}_{c_K}, \boldsymbol{\alpha}, \boldsymbol{\beta})\right] \tag{6.56}$$

$$\nabla_{\boldsymbol{\beta}} \mathbf{z}_c^T(\boldsymbol{\alpha}, \boldsymbol{\beta}) = \left[\nabla_{\boldsymbol{\beta}} \mathbf{z}^T(\mathbf{x}_{c_1}, \boldsymbol{\alpha}, \boldsymbol{\beta}), \dots, \nabla_{\boldsymbol{\beta}} \mathbf{z}^T(\mathbf{x}_{c_K}, \boldsymbol{\alpha}, \boldsymbol{\beta})\right] \tag{6.57}$$

where $\nabla_{\boldsymbol{\alpha}} \mathbf{z}^T(\boldsymbol{\vartheta})$ and $\nabla_{\boldsymbol{\beta}} \mathbf{z}^T(\boldsymbol{\vartheta})$ follow one of the sets of definitions in Table 6.2, and the measurement function follows from those in (6.8–6.11).

From [17], we take the inspiration for separation of the problem, where they note that the likelihood function for calibration measurements is separable, in that the gradient with respect to sensor position errors is independent of measurement bias. Therefore, the two types of errors can be separately estimated. In [17], they derive first an iterative approximation algorithm for measurement biases (under the assumption that sensor position estimates are accurate), then use those measurement bias estimates to solve for an estimate of sensor position error. We present a similar approach, but do not strictly follow their derivation or algorithms. Our two-stage algorithm is depicted graphically in Figure 6.11.

6.3.1 Measurement Bias Estimation

We begin by estimating measurement biases, under the assumption that the sensor positions are accurate ($\boldsymbol{\beta} = \boldsymbol{\beta}^0$). In this case, the likelihood function and the gradient of the likelihood (with respect to $\boldsymbol{\alpha}$) are given:

$$\ell(\boldsymbol{\zeta}_c|\boldsymbol{\alpha}) = \left[\boldsymbol{\zeta}_c - \mathbf{z}_c\left(\boldsymbol{\alpha}, \boldsymbol{\beta}^0\right)\right]^T \mathbf{C}_{\zeta_c}^{-1} \left[\boldsymbol{\zeta}_c - \mathbf{z}_c\left(\boldsymbol{\alpha}, \boldsymbol{\beta}^0\right)\right] \tag{6.58}$$

$$\nabla_{\boldsymbol{\alpha}} \ell(\boldsymbol{\zeta}_c|\boldsymbol{\alpha}) = \nabla_{\boldsymbol{\alpha}} \mathbf{z}_c^T\left(\boldsymbol{\alpha}, \boldsymbol{\beta}^0\right) \mathbf{C}_{\zeta_c}^{-1} \left[\boldsymbol{\zeta}_c - \mathbf{z}_c\left(\boldsymbol{\alpha}, \boldsymbol{\beta}^0\right)\right] \tag{6.59}$$

$$\nabla_{\boldsymbol{\alpha}} \mathbf{z}_c^T(\boldsymbol{\alpha}) = \left[\nabla_{\boldsymbol{\alpha}} \mathbf{z}^T\left(\mathbf{x}_{c_1}, \boldsymbol{\alpha}, \boldsymbol{\beta}^0\right), \dots, \nabla_{\boldsymbol{\alpha}} \mathbf{z}^T\left(\mathbf{x}_{c_K}, \boldsymbol{\alpha}, \boldsymbol{\beta}^0\right)\right] \tag{6.60}$$

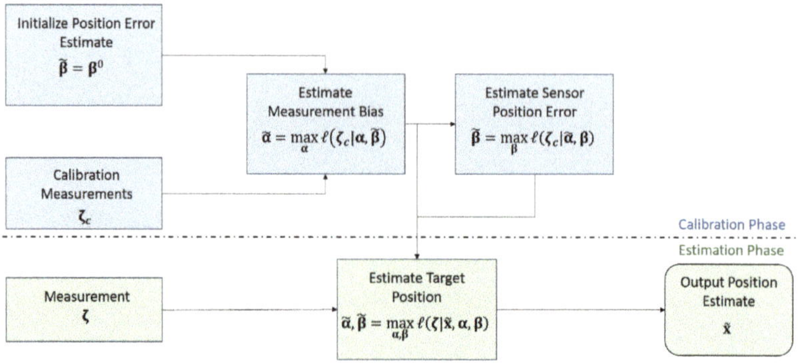

Figure 6.11 Illustration of the generic two-stage algorithm with calibration emitters.

where the gradients $\nabla_{\alpha}\mathbf{z}^T(\boldsymbol{\vartheta})$ follow those defined in Table 6.2.

We can, thus, complete this step using any of the defined solvers (`mlSoln`, `lsSoln`, or `gdSoln`), but we will need to pass in the calibration measurement function from (6.53), with the sensor position errors fixed at the reported positions $\boldsymbol{\beta} = \boldsymbol{\beta}^0$, and the bias vector $\boldsymbol{\alpha}$ as the input to estimate, along with the calibration covariance matrix $\mathbf{C}_{\zeta_c}$, and Jacobian matrix (6.56). The resulting bias estimates are noted $\widehat{\boldsymbol{\alpha}}$.

We have implemented this in the functions `mlSolnCal`, `lsSolnCal`, and `gdSolnCal`, in each of the `triang`, `tdoa`, `fdoa`, and `hybrid` packages. For example, this is achieved in the `triang.mlSolnCal` function with the lines:

```
theta = @(alpha) cat(1,x_cal(:),alpha,x_aoa(:));
ell = @(alpha) triang.loglikelihoodUnc(x_aoa, zeta_cal, C, C_beta, theta(
    alpha));
alpha_est = utils.mlSoln(ell_a,alpha_ctr,alpha_srch,eps_a);
```

The first line creates a function handle to construct the parameter vector $\boldsymbol{\vartheta}$, using the known calibration emitter and sensor positions. The second builds the likelihood function $\ell(\zeta_c|\boldsymbol{\alpha})$, and the estimate $\widehat{\boldsymbol{\alpha}}$ is calculated with a call to the standard ML estimator. The variables `alpha_ctr`, `alpha_srch` and `eps_a` define the search grid for the vector $\boldsymbol{\alpha}$.

There is one significant modification to `loglikelihoodUnc` that was done for this. Normally, the vector $\boldsymbol{\vartheta}$ is assumed to have only a single target position at its start, whereas we may begin with more than one calibration emitter. In order to handle this, we modified `loglikelihoodUnc` to examine the shape of the measurement vector. In this case, `zeta_cal` is a matrix, and each column corresponds to a different calibration emitter. In this manner, the likelihood function is able to parse the correct number of emitter positions from $\boldsymbol{\vartheta}$ and computes the cumulative likelihood across them all. This same behavior could be exploited for simultaneous multitarget position estimation, provided the measurements are properly sorted and the number of targets is known.

6.3.2 Sensor Position Error Estimation

Next, we use the bias estimates $\widehat{\alpha}$ and solve for estimated sensor positions and velocities β, which we will denote $\widehat{\beta}$. In this case, the likelihood function and the gradient with respect to sensor position are:

$$\begin{aligned}\ell\left(\zeta_c|\beta\right) &= -\frac{1}{2}\left(\widetilde{\zeta}-\widetilde{\mathbf{z}}(\mathbf{x},\widehat{\alpha},\beta)\right)^T\mathbf{C}_{\widetilde{\zeta}}^{-1}\left(\widetilde{\zeta}-\widetilde{\mathbf{z}}(\mathbf{x},\widehat{\alpha},\beta)\right)\\ &\quad -\frac{1}{2}\left(\beta-\beta^0\right)^T\mathbf{C}_\beta\left(\beta-\beta^0\right) \qquad (6.61)\\ \nabla_\beta\ell\left(\zeta_c|\beta\right) &= \nabla_\beta\mathbf{z}_c^T\left(\widehat{\alpha},\beta\right)\mathbf{C}_{\zeta_c}^{-1}\left[\zeta_c-\mathbf{z}_c\left(\widehat{\alpha},\beta\right)\right]-\mathbf{C}_\beta^{-1}\left(\beta-\beta^0\right) \qquad (6.62)\\ \nabla_\beta\mathbf{z}_c^T\left(\widehat{\alpha},\beta\right) &= \left[\nabla_\beta\mathbf{z}^T\left(\mathbf{x}_{c_1},\widehat{\alpha},\beta\right),\dots,\nabla_\beta\mathbf{z}^T\left(\mathbf{x}_{c_K},\widehat{\alpha},\beta\right)\right] \qquad (6.63)\end{aligned}$$

where the gradients $\nabla_\beta\mathbf{z}^T(\vartheta)$ follow those defined in Table 6.2.

Using these equations, we can solve for the estimate $\widehat{\beta}$ given the prior estimate of $\widehat{\alpha}$. In the case of the ML solver, this is done the same as in the previous section, except now we build the vector ϑ using the estimated bias $\widehat{\alpha}$ in addition to the calibration emitter positions, and we allow the solver to search for β.

```
theta_b = @(beta) cat(1,x_cal(:),alpha_est,beta);
ell_b = @(beta) triang.loglikelihoodUnc(x_aoa, zeta_cal, C, C_beta, theta_b(
    beta));

beta_est = utils.mlSoln(ell_b,beta_ctr,beta_srch,eps_b);
```

Once again, the first line creates our vector ϑ, the second line builds the likelihood function—this time it is $\ell\left(\zeta_c|\beta\right)$—and the estimate $\widehat{\beta}$ is computed with the ML solver. The terms `beta_ctr`, `beta_srch`, and `eps_b` define the search grid for β.

6.3.3 Estimating Target Position

The sensor uncertainty estimates $\widehat{\alpha}$ and $\widehat{\beta}$ that were generated from the calibration measurements ζ_c can then be passed on to any source localization algorithm to estimate the source position $\mathbf{x}$ from a set of measurements ζ, under the assumption that the biases α and sensor positions β match those used to generate the calibration data ζ_c. This is reasonable if the calibration measurements are gathered at the same time as the target signal measurements, or at approximately the same time.

Returning to our ML estimator example, this final stage is similarly built, except now that we are back to estimating the unknown target position $\mathbf{x}$ and using the measurements ζ instead of the calibration emitters and calibration measurements.

```
theta_x = @(x) cat(1,x,alpha_est,beta_est);
ell = @(x) triang.loglikelihoodUnc(x_aoa, zeta, C, C_beta, theta_x(x));
```

```
x_est = utils.mlSoln(ell,x_ctr,x_srch,eps_x);
```

Example 6.5 TDOA/FDOA with Calibration Emitter

Consider a 3-sensor joint TDOA/FDOA system with sensor velocity uncertainty (but no position uncertainty). Sensors are located at (−1, 0), (0, 1) km, and (1, 0) km, with reported velocity of (0, 500) m/s. Assume that the velocity uncertainty for all three sensors has covariance $\mathbf{C} = \sigma^2\mathbf{I}$, $\sigma^2 = (100m/s)^2$, generate a set of random velocities based on this distribution.

Generate signals for a stationary target at (−3, 4) km and a set of five calibration emitters evenly spaced between (2, −5) km and (2, −5) km, using measurement noise of $\sigma_t^2 = (100 \text{ ns})^2$ for TDOA and $\sigma_f^2 = (1 \text{ Hz})^2$ for FDOA, with a carrier of $f_0 = 10$ GHz.

Use the calibration signal to estimate sensor velocities, then use those velocities to estimate the target's position using a gradient descent solver, with an initial target estimate of (0, 3) km.

Solution: We begin by defining the sensor positions and nominal velocities, `x_tdoa`, `x_fdoa`, and `v_fdoa` (code omitted for brevity). Next, we generate velocity perturbations and use them to define the actual sensor velocities `v_fdoa_actual`.

```
C_vel = 100^2*eye(n_dim*n_fdoa);
U_vel = chol(C_vel,'upper');
vel_err = reshape(U_vel * randn(n_dim*n_fdoa,1),n_dim,n_fdoa);

v_fdoa_actual = v_fdoa + vel_err;
```

Next, we define the target position `x_tgt` and ideal measurements `z`, and the calibration emitter positions `x_cal` and ideal calibration measurements `z_cal`.

```
x_tgt = [-3;4] * 1e3;
n_cal = 5;
x_cal = [linspace(-2,2,5);-5*ones(1,n_cal)] * 1e3;

z = hybrid.measurement([],x_tdoa,x_fdoa,v_fdoa_actual,x_tgt, n_tdoa, n_fdoa)
    ;
z_cal = hybrid.measurement([],x_tdoa,x_fdoa,v_fdoa_actual,x_cal, n_tdoa,
    n_fdoa); % free of pos unc and bias
```

Before we generate noise for the measurements, we'll construct the sensor-level covariance matrix `C_tf`, which will be used when calling the solver functions. To do so, we first define the TDOA and FDOA sensor-level covariance matrices, in terms of range and range-rate, `C_roa` and `C_rroa`. Then, we concatenate the matrices diagonally.

```
% Build sensor-level covariance matrix
err_toa = 100e-9;
err_foa = 100;
f0 = 10e9;
lam = utils.constants.c/f0;
C_toa = err_toa^2*eye(n_tdoa);
C_roa = utils.constants.c^2*C_toa; % range error covariance
C_foa = err_foa^2*eye(n_fdoa);
C_rroa = lam^2*C_foa; % range rate error covariance
C_tf = blkdiag(C_roa, C_rroa); % assume TDOA/FDOA msmts are ind.
```

Now, to generate noise, we'll need measurement-level covariance matrices, expressed for the range difference and range-rate difference measurements. We construct these matrices independently as `C_rdoa` and `C_rrdoa`, respectively. Then, we generate noise for each independently, and then vertically stack them to make a complete set of noise measurements. Note that we generate `1+n_cal` columns of each. The first column will be for the target measurements, while the remaining `n_cal` columns are for the calibration measurements. This results in the vector `zeta` of noisy target measurements, and the matrix `zeta_cal` of calibration measurements for each of the `n_cal` calibration emitters.

```
% Generate noise; we'll do it separately for TDOA and FDOA
C_rdoa = utils.resampleCovMtx(C_roa,n_tdoa);
L_rdoa= chol(C_rdoa,'lower');
noise_rdoa = L_rdoa*randn(size(L_rdoa,2),1+n_cal);

C_rrdoa = utils.resampleCovMtx(C_rroa,n_tdoa);
L_rrdoa= chol(C_rrdoa,'lower');
noise_rrdoa = L_rrdoa*randn(size(L_rrdoa,2),1+n_cal);

noise = cat(1,noise_rdoa,noise_rrdoa);
zeta = z + noise(:,1); % noisy target measurements
zeta_cal = z_cal + noise(:,2:end); % noisy calibration measurements
```

To solve for position, we define the initial target position estimate. First, for comparison's sake, we perform an initial estimate using `gdSoln`, which assumes that the nominal position and velocity are accurate. The result is stored in `x_est`. Next, we use the calibration solver `gdSolnCal` to first use the calibration signals to estimate α and β as shown in Figure 6.11, then to estimate target position with calibrated sensor error estimates. This is stored in `x_est_cal`.

```
%% Estimate Position
x_init = [0; 3]*1e3;

[x_est,x_est_full] = hybrid.gdSoln([],x_tdoa,x_fdoa,v_fdoa,zeta,C_tf,x_init
    ,[],[],[],[],false,false,n_tdoa,n_fdoa);
```

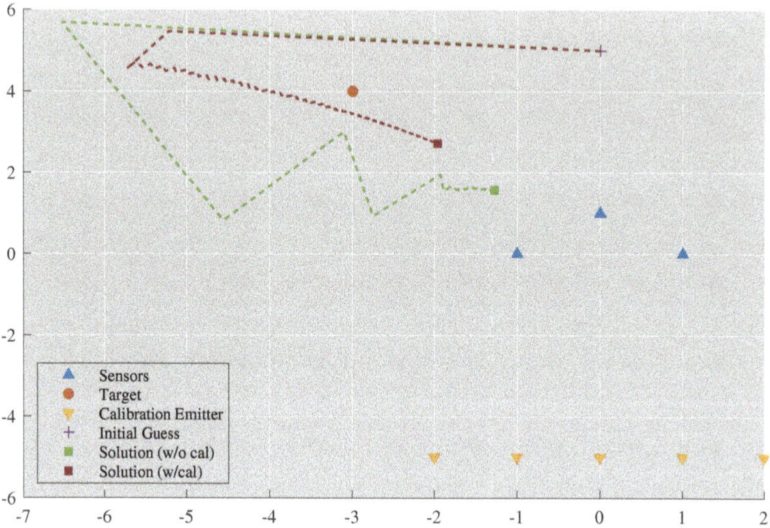

Figure 6.12 Sensor, calibration emitter, target, and estimated target positions for Example 6.5.

```
[x_est_cal,x_est_cal_full,~,~] = hybrid.gdSolnCal([],x_tdoa,x_fdoa,v_fdoa,
    zeta,x_cal,zeta_cal,C_tf,x_init,[],[],[],[],false,false,n_tdoa,n_fdoa);
```

The scenario, including the position of the TDOA/FDOA sensors, the calibration emitters, and both the raw and calibrated geolocation estimates, are shown in Figure 6.12. We observe that the calibration emitters enable the solution to more closely approximate the true target position in this case.

□

6.4 OTHER SOLUTION APPROACHES

In addition to the generic approaches that we have described, many scenario-specific solutions exist in the literature, with varying features, optimizations, and constraints. For instance, Jean and Weiss use a simulated annealing process, which mimics a metallurgical process used to strengthen alloys, to iteratively estimate the position of multiple sources using AOA with unknown measurement biases [11]. In another study, Montminy considers TDOA in an urban scenario, and uses an iterative approach to detect sensors with non line of sight conditions (which would violate TDOA measurement assumptions), remove their measurements, and geolocate targets using only well-behaved (line-of-sight) measurements [23].

In [24], Yang and Ho consider TDOA for multiple sources with sensor position error and derive a closed-form solution that relies on first hypothesizing locations for

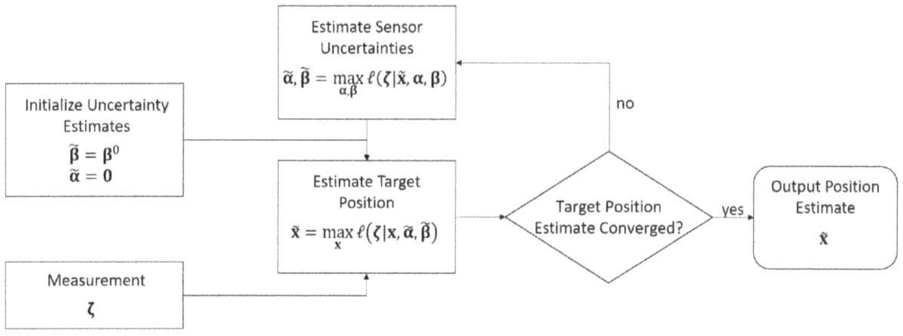

Figure 6.13 Illustration of a typical iterative adaptive solution algorithm, whereby the two sets of parameters (in this case, target position and sensor uncertainties) are solved one at a time, each using the most recent estimate of the other. The process is repeated until some convergence or stopping condition is met.

the two sources, then estimating true sensor position before ultimately reestimating source locations. Later, Wang and Ho [25] solve for source position directly, without estimating clock drift or sensor position errors first. Both of these closed form solutions are ultimately derived from the TDOA solution for an ideal (perfect sensor position) case in [26].

One possible extension of the separate estimation approach of [24], where sensor errors are first estimated with a hypothesized source position, which is then updated with the resultant sensor position estimates, is to iteratively repeat that process, until convergence is achieved. This is a general class of algorithm referred to as *iterative adaptive*,[9] and we illustrate one possible approach in Figure 6.13. In this case, an initial guess is made that the biases are zero and the sensor positions are known perfectly. Under these ideal conditions, we estimate the target position $\widehat{\mathbf{x}}$, and then that estimate is used to update the sensor position and measurement bias estimates, $\widehat{\boldsymbol{\beta}}$ and $\widehat{\boldsymbol{\alpha}}$, respectively. This process then repeats, iteratively updating each set of estimates in turn, until some stopping condition is reached, such as a maximum number of iterations or a convergence test on $\widehat{\mathbf{x}}$. This approach can work very well, particularly when the biases and sensor position errors are small, such that the likelihood function $\ell\left(\boldsymbol{\zeta}|\boldsymbol{\vartheta}\right)$ is locally convex between the true sensor parameter values $(\boldsymbol{\alpha}, \boldsymbol{\beta})$ and the initial assumed values $\left(\boldsymbol{\alpha} = \mathbf{0}, \boldsymbol{\beta} = \boldsymbol{\beta}^0\right)$.

The importance of calibrating sensors to the same reference orientation is briefly discussed in [30].

Tracking can also be an important part of overcoming sensor uncertainties, as the repeated measurements over time can improve estimates of slowly varying or static errors, such as in [31].

9. Iterative adaptive algorithms are found in a variety of estimation problems, such as array-based geolocation [27], radar imaging [28], and spectral estimation [29].

6.5 PERFORMANCE ANALYSIS

As noted before, we will use the *hybrid* CRLB, which essentially follows the same approach as the standard CRLB, but is applied to both the random uncertainties ($\boldsymbol{\beta}$) as well as to the deterministic unknowns ($\boldsymbol{\alpha}$). In this section, we'll use the shorthand $\boldsymbol{\vartheta}_n = \left[\boldsymbol{\alpha}^T, \boldsymbol{\beta}^T\right]^T$ to refer to the two sets of unknowns as *nuisance parameters*, since they must be estimated but are ultimately not important to system performance, except insofar as they impact accuracy of the target position estimate $\widehat{\mathbf{x}}$.

$$\mathbf{F}\left(\mathbf{x}, \boldsymbol{\vartheta}_n\right) = \begin{bmatrix} \mathbf{F}_{\mathbf{x}}\left(\mathbf{x}, \boldsymbol{\vartheta}_n\right) & \mathbf{F}_{\mathbf{x},\boldsymbol{\vartheta}_n}\left(\mathbf{x}, \boldsymbol{\vartheta}_n\right) \\ \mathbf{F}_{\mathbf{x},\boldsymbol{\vartheta}_n}^T\left(\mathbf{x}, \boldsymbol{\vartheta}_n\right) & \mathbf{F}_{\boldsymbol{\vartheta}_n}\left(\mathbf{x}, \boldsymbol{\vartheta}_n\right) \end{bmatrix} \tag{6.64}$$

From [25], [32], we can show that these submatrices are defined:

$$\mathbf{F}_{\mathbf{x}}\left(\mathbf{x}, \boldsymbol{\vartheta}_n\right) = \nabla_{\mathbf{x}}\mathbf{z}^T\left(\mathbf{x}, \boldsymbol{\vartheta}_n\right)\mathbf{C}_{\zeta}^{-1}\left[\nabla_{\mathbf{x}}\mathbf{z}^T\left(\mathbf{x}, \boldsymbol{\vartheta}_n\right)\right]^T \tag{6.65}$$

$$\mathbf{F}_{\mathbf{x},\boldsymbol{\vartheta}_n}\left(\mathbf{x}, \boldsymbol{\vartheta}_n\right) = \nabla_{\mathbf{x}}\mathbf{z}^T\left(\mathbf{x}, \boldsymbol{\vartheta}_n\right)\mathbf{C}_{\zeta}^{-1}\begin{bmatrix} \nabla_{\boldsymbol{\alpha}}\mathbf{z}^T\left(\mathbf{x}, \boldsymbol{\vartheta}_n\right) \\ \nabla_{\boldsymbol{\beta}}\mathbf{z}^T\left(\mathbf{x}, \boldsymbol{\vartheta}_n\right) \end{bmatrix}^T \tag{6.66}$$

$$\begin{aligned} F_{\boldsymbol{\vartheta}_n}\left(\mathbf{x}, \boldsymbol{\vartheta}_n\right) = &\begin{bmatrix} \nabla_{\boldsymbol{\alpha}}\mathbf{z}^T\left(\mathbf{x}, \boldsymbol{\vartheta}_n\right) \\ \nabla_{\boldsymbol{\beta}}\mathbf{z}^T\left(\mathbf{x}, \boldsymbol{\vartheta}_n\right) \end{bmatrix}\mathbf{C}_{\zeta}^{-1}\begin{bmatrix} \nabla_{\boldsymbol{\alpha}}\mathbf{z}^T\left(\mathbf{x}, \boldsymbol{\vartheta}_n\right) \\ \nabla_{\boldsymbol{\beta}}\mathbf{z}^T\left(\mathbf{x}, \boldsymbol{\vartheta}_n\right) \end{bmatrix}^T \\ &+ \begin{bmatrix} \mathbf{0} & \mathbf{0} \\ \mathbf{0} & \mathbf{C}_{\boldsymbol{\beta}}^{-1} \end{bmatrix} \end{aligned} \tag{6.67}$$

The CRLB for the sensor position is the upper left quadrant of the inverse of $\mathbf{F}$, which we can compute via the Schur complement [12]:

$$\mathbf{C}_{\epsilon}(\mathbf{x}) \geq \left[\mathbf{F}_{\mathbf{x}} - \mathbf{F}_{\mathbf{x},\mathbf{a}}\mathbf{F}_{\mathbf{a}}^{-1}\mathbf{F}_{\mathbf{x},\mathbf{a}}^T\right]^{-1} \tag{6.68}$$

In the absence of sensor parameters errors, the latter terms in (6.68) will fall out, and the CRLB reduces to the forms used in Chapter 2.

6.6 PROBLEM SET

6.1 Consider Example 6.1, but add a calibration emitter at (0, 2). Use that to estimate measurement bias, then estimate the target position.

6.2 Consider Example 6.2, but add a calibration emitter at (2, 0). Use that to estimate sensor positions, then estimate the target position.

6.3 Consider Example 6.3, but add a calibration emitter at (0, −2) and another at (0, 3). Use these calibration emitters to estimate sensor position and measurement biases, then estimate the target position.

6.4 Repeat Example 6.5, but solve for target position with maximum likelihood, using a grid with .1-km precision in x and y, for a search grid spanning $(-5, 3)$ and (5, 10) km. For calibration, search for the true velocity with 1-m/s precision in v_x and v_y, spanning ± 150 m/s in both v_x and v_y.

References

[1] N. A. O'Donoughue, *Emitter Detection and Geolocation for Electronic Warfare*. Norwood, MA: Artech House, 2019.

[2] R. Bardelli, D. Haworth, and N. Smith, "Interference localisation for the eutelsat satellite system," *Proc. GLOBECOM'95*, vol. 3, pp. 1641–1651, 1995.

[3] N. E. Gemayel, H. Jäkel, and F. K. Jondral, "Error analysis of a low cost TDOA sensor network," *Proc. of IEEE/ION PLANS 2014*, pp. 1140–1145, 2014.

[4] T. E. Humphreys, B. M. Ledvina, M. L. Psiaki, B. W. O'Hanlon, P. M. Kintner, *et al.*, "Assessing the spoofing threat: Development of a portable GPS civilian spoofer," *Proc. of the 21st International Technical Meeting of the Satellite Division of The Institute of Navigation (ION GNSS 2008)*, pp. 2314–2325, 2008.

[5] N. O. Tippenhauer, C. Pöpper, K. B. Rasmussen, and S. Capkun, "On the requirements for successful GPS spoofing attacks," *Proc. of the 18th ACM Conference on Computer and Communications Security*, pp. 75–86, 2011.

[6] T. Humphreys, "Statement on the vulnerability of civil unmanned aerial vehicles and other systems to civil GPS spoofing," *Submitted to the Subcommittee on Oversight, Investigations, and Management of the House Committee on Homeland Security*, p. 16, 2012.

[7] L. Landesa, I. Castro, J. Taboada, and F. Obelleiro, "Bias of the maximum likelihood doa estimation from inaccurate knowledge of the antenna array response," *Journal of Electromagnetic Waves and Applications*, vol. 21, no. 9, pp. 1205–1217, 2007.

[8] Z. Liu, Z. Huang, and Y. Zhou, "Bias analysis of music in the presence of mutual coupling," *IET Signal Processing*, vol. 3, 74–84(10), 1 Jan. 2009, ISSN: 1751-9675. [Online]. Available: `https://digital-library.theiet.org/content/journals/10.1049/iet-spr_20070213`.

[9] H. Messer, "The Hybrid Cramer-Rao Lower Bound - from practice to theory," in *Fourth IEEE Workshop on Sensor Array and Multichannel Processing, 2006.*, 2006, pp. 304–307. DOI: `10.1109/SAM.2006.1706142`.

[10] Z. Ma and K. C. Ho, "A study on the effects of sensor position error and the placement of calibration emitter for source localization," *IEEE Transactions on Wireless Communications*, vol. 13, no. 10, pp. 5440–5452, 2014. DOI: `10.1109/TWC.2014.2341609`.

[11] O. Jean and A. J. Weiss, "Geolocation by direction of arrival using arrays with unknown orientation," *IEEE transactions on signal processing*, vol. 62, no. 12, pp. 3135–3142, 2014.

[12] L. L. Scharf, *Statistical Signal Processing*. Reading, MA: Addison-Wesley, 1991.

[13] A. Kintz and I. J. Gupta, "Airborne antenna array calibration with signals of opportunity," in *2013 IEEE Antennas and Propagation Society International Symposium (APSURSI)*, 2013, pp. 1264–1265. DOI: `10.1109/APS.2013.6711292`.

[14] F. Pang, K. Doğançay, N. H. Nguyen, and Q. Zhang, "Aoa pseudolinear target motion analysis in the presence of sensor location errors," *IEEE Transactions on Signal Processing*, vol. 68, pp. 3385–3399, 2020. DOI: `10.1109/TSP.2020.2998896`.

[15] K. Ho and L. Yang, "On the use of a calibration emitter for source localization in the presence of sensor position uncertainty," *IEEE Transactions on Signal Processing*, vol. 56, no. 12, pp. 5758–5772, 2008.

[16] L. Yang and K. Ho, "Alleviating sensor position error in source localization using calibration emitters at inaccurate locations," *IEEE transactions on signal processing*, vol. 58, no. 1, pp. 67–83, 2009.

[17] D. Wang, J. Yin, X. Chen, C. Jia, and F. Wei, "On the use of calibration emitters for TDOA source localization in the presence of synchronization clock bias and sensor location errors," *EURASIP Journal on Advances in Signal Processing*, vol. 2019, no. 1, pp. 1–34, 2019.

[18] K. Ren and Z. Sun, "Optimum strategy of reference emitter placement for dual-satellite TDOA and FDOA localization," *Prof. of Eighth International Conference on Instrumentation Measurement, Computer, Communication and Control (IMCCC 2018)*, pp. 474–478, 2018. DOI: 10.1109/IMCCC.2018.00105.

[19] J. Li, F. Guo, and W. Jiang, "Source localization and calibration using TDOA and FDOA measurements in the presence of sensor location uncertainty," *Science China Information Sciences*, vol. 57, no. 4, pp. 1–12, 2014.

[20] L. Zhang, D. Wang, and Y. Wu, "Performance analysis of TDOA and FDOA location by differential calibration with calibration sources.," *J. Commun.*, vol. 9, no. 6, pp. 483–489, 2014.

[21] D. Wang, P. Zhang, Z. Yang, F. Wei, and C. Wang, "A novel estimator for TDOA and FDOA positioning of multiple disjoint sources in the presence of calibration emitters," *IEEE Access*, vol. 8, pp. 1613–1643, 2020. DOI: 10.1109/ACCESS.2019.2959235.

[22] T. Pattison and S. Chou, "Sensitivity analysis of dual-satellite geolocation," *IEEE Transactions on Aerospace and Electronic Systems*, vol. 36, no. 1, pp. 56–71, 2000.

[23] M. B. Montminy, "Passive geolocation of low power emitters in urban environments using TDOA," M.S. thesis, Air Force Institute of Technology, 2007.

[24] L. Yang and K. C. Ho, "An approximately efficient TDOA localization algorithm in closed-form for locating multiple disjoint sources with erroneous sensor positions," *IEEE Transactions on Signal Processing*, vol. 57, no. 12, pp. 4598–4615, Dec. 2009, ISSN: 1053-587X. DOI: 10.1109/TSP.2009.2027765.

[25] Y. Wang and K. Ho, "TDOA source localization in the presence of synchronization clock bias and sensor position errors," *IEEE Transactions on Signal Processing*, vol. 61, no. 18, pp. 4532–4544, 2013.

[26] Y.-T. Chan and K. Ho, "A simple and efficient estimator for hyperbolic location," *IEEE Transactions on Signal Processing*, vol. 42, no. 8, pp. 1905–1915, 1994.

[27] T. Yardibi, J. Li, P. Stoica, M. Xue, and A. B. Baggeroer, "Source localization and sensing: A nonparametric iterative adaptive approach based on weighted least squares," *IEEE Transactions on Aerospace and Electronic Systems*, vol. 46, no. 1, pp. 425–443, 2010. DOI: 10.1109/TAES.2010.5417172.

[28] W. Roberts, P. Stoica, J. Li, T. Yardibi, and F. A. Sadjadi, "Iterative adaptive approaches to mimo radar imaging," *IEEE Journal of Selected Topics in Signal Processing*, vol. 4, no. 1, pp. 5–20, 2010. DOI: 10.1109/JSTSP.2009.2038964.

[29] G.-O. Glentis and A. Jakobsson, "Efficient implementation of iterative adaptive approach spectral estimation techniques," *IEEE Transactions on Signal Processing*, vol. 59, no. 9, pp. 4154–4167, 2011. DOI: 10.1109/TSP.2011.2145376.

[30] D. Adamy, *EW 103: Tactical Battlefield Communications Electronic Warfare*. Norwood, MA: Artech House, 2008.

[31] S.-K. Han, W.-S. Ra, and J. B. Park, "TDOA/FDOA based target tracking with imperfect position and velocity data of distributed moving sensors," *International Journal of Control, Automation and Systems*, vol. 15, no. 3, pp. 1155–1166, 2017.

[32] K. C. Ho, X. Lu, and L. Kovavisaruch, "Source Localization Using TDOA and FDOA Measurements in the Presence of Receiver Location Errors: Analysis and Solution," *IEEE Transactions on Signal Processing*, vol. 55, no. 2, pp. 684–696, Feb. 2007, ISSN: 1053-587X. DOI: 10.1109/TSP.2006.885744.

Chapter 7

Geolocation with Multiple Snapshots

Until this point, we have assumed the use of a single snapshot ζ, made up of some combination of TDOA, FDOA, and AOA measurements, sometimes supplemented with calibration emitters, but only ever collected at a single point in time, to estimate the position of the target. In this chapter, we consider the use of repeated measurements ζ_k, $k = 0, \dots, K-1$.

Additional measurements are primarily useful for two reasons. First, additional measurements result in reduced estimation error, leading ultimately to reduced geolocation error, provided that the measurements follow the same distribution and are uncorrelated with each other (in other words, that they are *independent and identically distributed*—IID)[1].

The second possible benefit is that repeated measurements, again under certain conditions, can improve the *observability* of the problem and provide sufficient new information to accurately estimate all of the unknowns. This can happen when there are sensors unknowns (as discussed in Chapter 6), or when there is degenerate geometry (provided that the sensors are moving with each snapshot, in order to provide the geometric diversity that they lacked in a single snapshot).

For example, in [2], we observed many geometries where there was a poor arrangement of sensors, such as when two AOA sensors formed a straight line to the target, resulting in a redundant measurement, or when the baseline between two TDOA sensors was nearly parallel to the line of sight from one to the target, resulting in a *geometric dilution of precision* that increased geolocation error significantly.

We'll examine these two potential benefits of repeated measurements in this chapter.

7.1 CONVERGENCE OF MULTIPLE MEASUREMENTS

Consider the collection of K snapshots, ζ_k, $k = 0, \dots, K-1$. Each of these measurements is assumed to be independent and identically distributed as a Gaussian random vector:[1].

$$\zeta_k \sim \mathcal{N}\left(\mathbf{z}(\mathbf{x}), \mathbf{C}_\zeta\right) \tag{7.1}$$

The most straightforward approach to using all of the measurements is to compute the sample mean, denoted $\overline{\zeta}_k$:

$$\overline{\zeta} = \frac{1}{K}\sum_{k=0}^{K-1} \zeta_k \sim \mathcal{N}\left(\mathbf{z}(\mathbf{x}), \frac{1}{K}\mathbf{C}_\zeta\right) \tag{7.2}$$

We can see that the expectation of $\overline{\zeta}$ is unchanged, but that its covariance matrix has been reduced by a factor $1/K$. Thus, we can use the same likelihood approach and solutions as we've done before in single-measurement cases, and expect roughly $1/K$ improvement in the error variance of both the measurements and the eventual geolocation estimates.

Example 7.1 Repeated AOA Measurements

Consider two AOA sensors at (0, 0) and (1, 0) km in a 2-D scenario, with $\sigma_\theta = 5°$ as the standard deviation on measurement accuracy for both sensors.[2] What is the CEP_{50} based on the CRLB for geolocation of an emitter at (2, 4) km?

How many measurements are required to achieve a CEP_{50} of 100 m?

Plot the CEP_{50}, based on the CRLB, for positions between 1 and 5 km downrange, and -5 to 5 km cross-range, with K=10 measurements and again for K=100 measurements. Discuss.

Solution: After defining the sensor positions (`x_aoa`), target position (`x_tgt`), and sensor accuracy (`sigma_psi`), we construct the measurement covariance matrix, use it to compute the CRLB for a single measurement, and use that to compute a lower bound on CEP_{50}, which results in 1.65 km.

```
n_sensors = size(x_aoa,2);
C = sigma_psi^2 * eye(n_sensors);

crlb = triang.computeCRLB(x_aoa, x_tgt, C);
```

Next, we compute how many samples it would require to achieve a CEP of 100 m. Note that the covariance matrix scales with $1/K$, and the CEP with $\mathbf{C}^{1/2}$, and so we expect the CEP to scale with $1/\sqrt{K}$. Thus, to lower the CEP from 1.65 km to 100 m, we'd expect a little over 100 samples. Doing the math results in 274 samples required.

1. Recall from [2] that none of the measurements are actually Gaussian, but that this is a convenient approximation that we make in order to generate closed-form estimators and performance predictions.
2. See Chapters 7 and 8 of [2] for details on how to compute σ_θ given the received signal strength and the type of AOA system in use.

To test this, we compute the CEP directly for a set of sample mean inputs with up to 1,000 measurements:

```
num_samples = 1:1000;
cep_vec = zeros(size(num_samples));
for idx_n = 1:numel(num_samples)
    this_num_samples = num_samples(idx_n);
    this_C = C/this_num_samples;

    this_crlb = triang.computeCRLB(x_aoa, x_tgt, this_C);
    cep_vec(idx_n) = utils.computeCEP50(this_crlb);
end
```

The result is plotted in Figure 7.1, for a larger interval of 1 to 10,000 samples, along with dashed lines for the desired CEP and the predicted number of samples. In addition to the stated 5° error, we also plot curves for 10° and 30°, for illustration. Figure 7.1 confirms our estimate; 274 samples yields 100 m CEP_{50}.

To provide some context, let's assume that the pulsed transmission in question is a Link-16 transmitter. The Link-16 protocol includes a 6.4 μs pulse every 13 μs. These pulses are organized into burst transmissions that last approximately 3.3 ms, and are placed into 7.8125 ms long time slots (the additional time is to allow for propagation of the signals, and a preamble to synchronize receivers) [3]. If we ignore the preamble, that means each time slot will contain approximately 253 pulses. These time slots are collected into 12-second frames (each containing 1,536 time slots), and there are 64 frames, which repeat every 12.8 minutes. The allocation of time slots to transmitters within each frame is complex, and dependent on the specific role of each transmitter, as well as the active subnets operating in a Link-16 network. Nevertheless, the scenario provided here could represent a single measurement of transmitter position, averaged across one 7.8125-ms-long time slot, and repeated measurements across the frames and epochs could be strung together to form a track (which we will discuss in Chapter 8).

Next, we compute the CEP_{50} for an area of regard that covers -5 to 5 km in the x-axis and 1 to 5 km in the y-axis, with both $K = 10$ and $K = 100$. For our solution, we have chosen a grid of positions that has 101 rows and 101 columns.

```
n_pts = 101;
x_vec = linspace(-5e3,5e3,n_pts);
y_vec = linspace(1e3,5e3,n_pts);
[xx,yy] = meshgrid(x_vec,y_vec);
x_aor = [xx(:), yy(:)]';
```

Next, we simply call the `computeCEP` function with two difference measurement covariances, one for $K = 10$ and one for $K = 100$.

```
aor_dims = size(xx);
crlb_k10 = triang.computeCRLB(x_aoa, x_aor, C/10);
cep_k10 = reshape(utils.computeCEP50(crlb_k10), aor_dims);
```

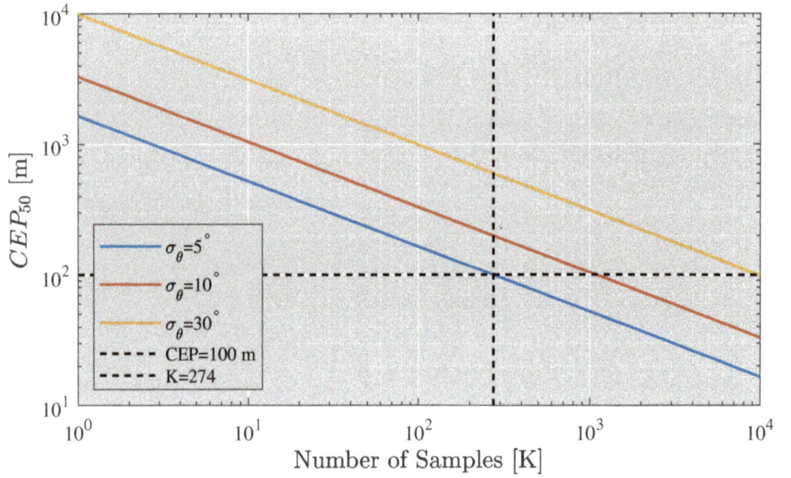

Figure 7.1 Chart of the CEP_{50}, as a function of the number of samples used to compute the sample mean, in Example 7.1.

```
crlb_k100 = triang.computeCRLB(x_aoa, x_aor, C/100);
cep_k100 = reshape(utils.computeCEP50(crlb_k100), aor_dims);
```

For simplicity, we omit the plotting commands.[3] The results are plotted in Figure 7.2. Figure 7.2(a) plots $K = 10$, while Figure 7.2(b) plots $K = 100$. We see identical performance *trends* in the two cases, but with a different scale to the errors. This, once again, confirms the relationship between geolocation error and $1/\sqrt{K}$. □

We now discuss how to leverage this convergence for geolocation using two approaches. In the first, which we'll call *measurement space solutions*, we will leverage the sample mean directly and perform geolocation using standard approaches. In the latter, which we'll call *state space solutions*, we will feed each measurement directly into a modified geolocation solver that will update the position estimate with new information as new measurements are provided.

A target's *state* refers to parameters, such as position and velocity and sometimes acceleration, but may include other parameters, such as orientation or configuration. The astute reader will notice that we are using language similar to that used in tracking problems. This is not an accident, as the type of problem considered here is a special case of tracking where the target's *state* is constant.

3. Careful readers will note that in the code provided for this example, we implemented the above section as a *for loop*, to allow for easy modification of the desired number of samples K and automatic plotting of each.

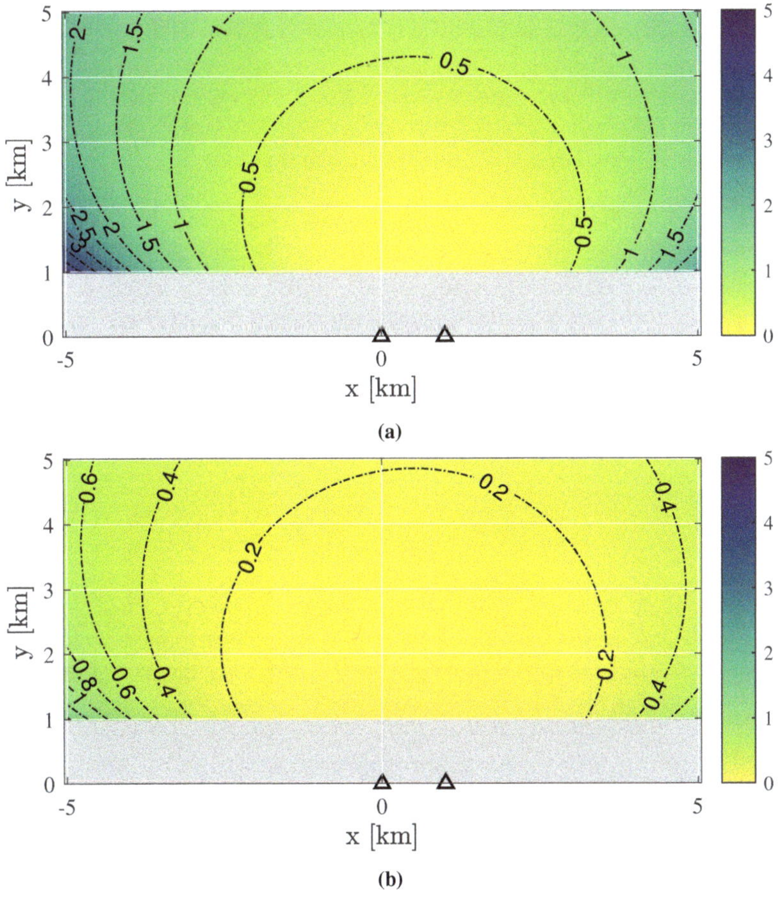

Figure 7.2 Geolocation error as computed by the CRLB for a two-sensor AOA scenario in Example 7.1. (a) CEP_{50} after $K = 10$ measurements, and (b) CEP_{50} after $K = 100$ measurements.

7.1.1 Measurement Space Solutions

In this first approach, our aim is to use the IID nature of the measurements ζ_k to improve the *inputs* to a geolocation solver, under the assumption that this will lead to improved *outputs*.

This is a reasonable assumption, since geolocation performance is directly dependent on measurement covariance (see Chapter 2 for derivations of geolocation error CRLB).

Geolocation with the sample mean follows the same approach as before; we simply replace the single measurement vector ζ with the sample mean $\overline{\zeta}_k$ when using any of the solvers from Section 2.2 or elsewhere in this text.

Example 7.2 TDOA Geolocation with Sample Mean

Consider a 3-sensor TDOA system at $(-10, 0)$, $(0, 10)$, and $(10, 0)$ km. A pulsed radar is operating at $(5, -15)$ km, with a pulse repetition interval of 1 ms. Each pulse is captured independently by the three sensors, with a timing accuracy of $\sigma_t = 1$ μs.[4] The sensors listen for a period of 1 second to refine their timing estimates. What is the expected CEP_{50} for geolocation accuracy, based on the CRLB?

How long must the sensors listen to achieve CEP < 10 m (again, based on the CRLB)?

Use the LS solver to demonstrate this problem with a 1 s interval. Generate noisy TDOA measurements, compute the sample mean, and then estimate the target position.

Solution: We begin, as usual, by defining the sensor positions (`x_tdoa`), target position (`x_tgt`) and measurement covariance matrix (`C_toa` for individual time measurements, `C_roa` for individual range measurements, and `C_rdoa` for range difference measurements). We calculate that a 1-second interval, with 1-ms PRI will yield 1,001 pulses.

```
pri = 1e-3;
T = 1; % observation period
num_pulses = floor(T/pri)+1;
```

Next, we compute the CRLB and CEP_{50} for both a single sample (using `C_toa`) and the sample mean (using `C_toa/num_pulses`). This yields 1.71 km for a single sample, and .05 km for the sample mean.

```
crlb_single_sample = tdoa.computeCRLB(x_tdoa, x_tgt, C_toa, ref_idx);
crlb_sample_mean = tdoa.computeCRLB(x_tdoa, x_tgt, C_toa/num_pulses, ref_idx
    );
cep_single_sample = utils.computeCEP50(crlb_single_sample);
cep_sample_mean = utils.computeCEP50(crlb_sample_mean);
```

4. See Chapter 11 of [2] for details on how to estimate σ_t for a given scenario and received signal strength.

To determine how long they must listen for $CEP_{50} < 10$ m, we could again do a back-of-the-envelope calculation that $K = 29,421$ samples (29.42 s given the 1-ms PRI) are needed, based on the square of 1.71 km / 10 m. We verify it with the following block of code that operates in 100-ms increments:

```
time_vec = 1:.1:100; % seconds
cep_vec = zeros(size(time_vec));
for idx_t = 1:numel(time_vec)
    this_time = time_vec(idx_t);
    this_num_samples = 1 + floor(this_time/pri);
    this_C = C_toa/this_num_samples;

    this_crlb = tdoa.computeCRLB(x_tdoa, x_tgt, this_C, ref_idx);
    cep_vec(idx_t) = utils.computeCEP50(this_crlb);
end
```

The result is 29.2 s, or roughly $K = 29,200$ measurements. It is unlikely that a threat would operate a stable waveform for this long, particularly an LPI radar, as we have hypothesized. If it is moving, or the sensors are moving, it will likely result in errors when we use that many pulses to estimate ζ.

The CEP_{50} is plotted as a function of time in Figure 7.3. For illustration purposes, we also show the CEP_{50} for the same scenario with $\sigma_t = 100$ ns (0.1 μs) and $\sigma_t = 10$ μs. We see that the latter could achieve the desired accuracy with approximately 30 ms (K=300).

This shows how the target signal of interest can be analyzed to determine what level of sensor performance is required to achieve a desired level of geolocation accuracy, given how long a pulsed target signal can be expected to operate coherently.

Next, we generate some noisy measurements, corresponding to this 30-ms interval ($K = 301$ pulses). The true measurement is generated once, and the range difference covariance matrix is used to generate random noise for each of the K measurements.

```
z = tdoa.measurement(x_tdoa, x_tgt, ref_idx);
L = chol(C_rdoa,'lower');
zeta = z + L*randn(size(C_rdoa,1),num_pulses); % noisy measurement
```

Then, we compute the running sample mean according to (7.2), and run the LS solver `tdoa.lsSoln` for each of the measurements (baseline) or sample mean calculations. The seed location from which each call to `lsSoln` begins is (1, 1) km.

```
zeta_mn = cumsum(zeta,2)./(1:num_pulses);

for idx=1:num_pulses
    [x_ls(:,idx), ~] = tdoa.lsSoln(x_tdoa, zeta(:,idx), C_roa, x_init);
    [x_ls_mn(:,idx), ~] = tdoa.lsSoln(x_tdoa, zeta_mn(:,idx), C_roa/idx,
        x_init);
end
```

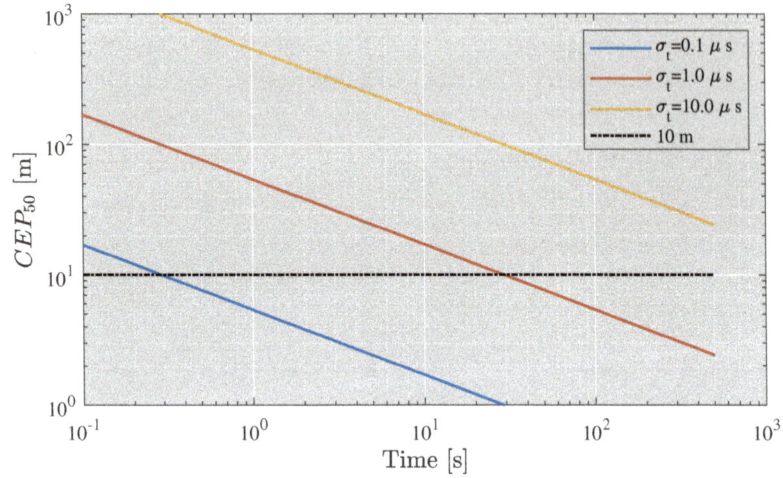

Figure 7.3 Plot of the expected CEP_{50}, as a function of collection interval, for Example 7.2.

The result is plotted in Figure 7.4(a), which shows the true target position as a blue triangle and the LS solution as a red dashed line. The CRLB for a single measurement geolocation solution is shown in the yellow ellipse (1σ confidence interval), and for the sample mean in the purple ellipse, which is too small to see in this chart. Figure 7.4(b) plots the error (euclidean distance) for each geolocation solution, as a function of time, as well as the CEP_{50} computed from the CRLB. The baseline (single-measurement geolocation) is plotted in blue, while the CRLB-derived CEP_{50} for that case is in yellow. The error from geolocation with the sample mean is shown in red, and its CRLB-derived CEP_{50} bound is in purple.

The first thing to notice is that while the single-sample error is stationary (sometimes performing quite well, and sometimes poorly, but well centered on the CRLB-derived bound), the sample-mean generated geolocation results generally get better over time. There is in interesting dip between .15 and .35 seconds, where it does considerably better than the bound, but as this is a single trial, it is likely an abnormality. It just so happened that for this trial, some of the errors lined up so that the sample mean was very close to the true measurement result, and it wasn't until .5 seconds that the errors built back up closer to what one would expect. By the .8-second mark, the achieved error is tracking quite well to the bound. □

7.1.1.1 Recursive Calculation of Sample Mean

The downside to this approach (computing the sample mean) is that it requires keeping all of the data in memory, and reevaluating both the sample mean every time a new measurement is received.

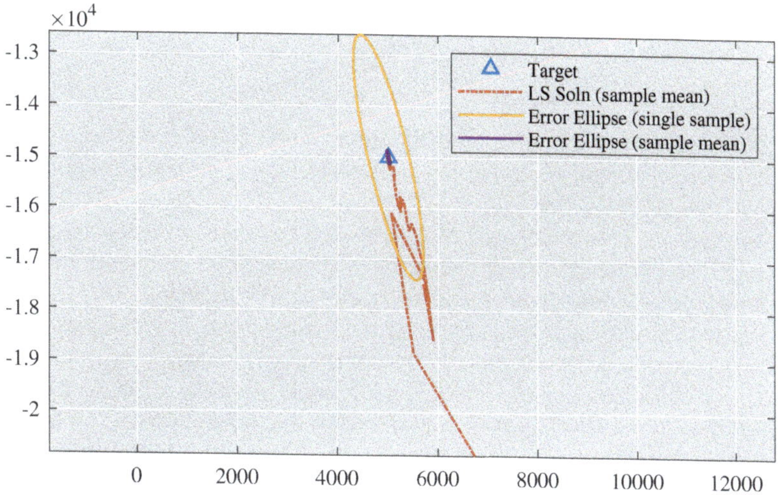

(a)

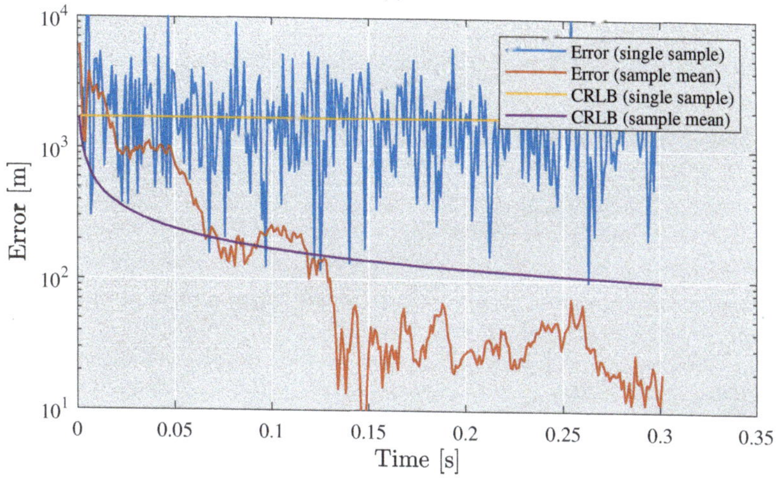

(b)

Figure 7.4 Geolocation results for the 3-sensor TDOA scenario in Example 7.2. (a) Layout of the target area, error ellipse (for both single-measurement geolocation, and sample-mean geolocation), and trajectory of sample-mean solutions, and (b) geolocation error, as a function of observation time, for single-measurement geolocation (blue) and sample mean-based geolocation (red), along with the CRLB for each case.

Equivalently, one could perform an update step on the sample mean, sometimes referred to as *sequential least square*[4].

$$\overline{\zeta}_k = \frac{k-1}{k}\overline{\zeta}_{k-1} + \frac{1}{k}\zeta_k \tag{7.3}$$

In this manner, we're computing the same sample mean, but at each step we need only the previous sample mean, the new measurement, and the number of measurements that have been taken thus far.

We have implemented a pair of utilities that can assist in calculating the sample mean, either as a batch process or via iterative updates. They are `computeSampleMean` and `computeSampleMeanUpdate`, and are in the `+utils` package.

Regardless of the manner of updating the sample mean, in both cases the resulting geolocation solution is derived directly from that sample mean.

This is statistically optimal, in that it considers all of the available measurements to minimize noise in the input data and solves for the target position [1]. It can also be computationally intense, particularly if the solver chosen is a brute force ML method and we execute that solver again with each new measurement ζ_k.

In Example 7.2, we execute geolocation once every ms. We could, perhaps, batch the measurements, and only update the position estimate once every 100 ms, or every 1 s, depending on algorithm complexity and available computing resources.

7.1.2 State Space Solutions

In state space solutions, the goal is to leverage the IID nature of the measurements *as they relate to the target's state* (in this case, position and velocity). In other words, we leverage the fact that just as the measurements (if averaged) converge to the true underlying parameter $\mathbf{z}(\mathbf{x})$, so to would a series of single-measurement position estimates (if averaged) converge to the true target position $\mathbf{x}$.[5]

The obvious approach is to do just that; average the position estimates over time. We leave this to the reader, as it is a straightforward implementation and carries many of the computational limitations as the measurement space solutions. Instead, we turn to a slightly more nuanced approach.

As an alternative, consider a recursive solver, such that we generate an estimate with the first snapshot $\widehat{\mathbf{x}}_0$, and then update that estimate with each successive measurement:

$$\widehat{\mathbf{x}}_0 = f\left(\zeta_o\right) \tag{7.4}$$

$$\widehat{\mathbf{x}}_k = f_R\left(\widehat{\mathbf{x}}_{k-1}, \zeta_k\right), \; k = 1, \ldots, K-1 \tag{7.5}$$

5. This also assumes the geolocation solution is unbiased; otherwise this average will converge to a biased solution.

where $f(\cdot)$ is one of the standard geolocation solvers, and $f_R(\cdot)$ is a recursive solver that updates its position estimate each time a new measurement is supplied. The question then remains, how do we design such a recursive solver?

A naive approach could use the previous estimate ($\widehat{\mathbf{x}}_{k-1}$) as a seed for an iterative solver like LS or GD, or to define the search space for a brute force solver, like ML. In this manner, each geolocation update is relying on the prior information to seed the solution, but then attempting to geolocate based solely on the new information (ζ_k). This is not optimal, since the result is only marginally influenced by the prior measurements, and the solution is principally a product only of the final measurement.

7.1.2.1 Kalman-Inspired Updates

Instead, we take inspiration from the ubiquitous Kalman filter [5], albeit greatly simplified, since the target is stationary and, in this case, so are the sensors. The Kalman filter is, at its heart, a sequential Bayes' estimator. At each measurement, Bayesian principles are used to take an a priori estimate of the target's state and a measurement to generate an a posterior estimate of the target's state[1]. Then, a model of the system is used to take that update and predict it forward to the next time instant, to be used as the a priori estimate when the next measurement arrives. This results in a two-stage *predict/update* cycle, which is a feature of many modern tracker formulations. In this chapter, there is no system dynamic, and so we focus only on the *update* portion of the tracker. We will discuss this in greater detail in Chapter 8, but interested readers are referred to one of the many fine tutorials and introductions to the Kalman Filter that can be found in the literature, such as [5]–[12].

At each update, we will perform four fundamental calculations that consider not only the previous position estimate and new measurement, but also the statistical distribution of each, to arrive at an optimal (under Gaussian conditions) combination of the available information.

The result will be an estimate of target position using data up until time k ($\mathbf{x}_k$) and the error covariance matrix of that estimate ($\mathbf{P}_k$).

1. *Compute the innovation (or residual).* With each new measurement ζ_k, we begin by computing the innovation, sometimes referred to as the residual, which is the difference between the observed measurement ζ_k and the measurement that we expected to observe:

$$\mathbf{y}_k = \zeta_k - \mathbf{z}\left(\mathbf{x}_{k-1}\right) \tag{7.6}$$

   where, as elsewhere in this text, $\mathbf{z}(\mathbf{x})$ is the measurement function.

   The residual also features heavily in iterative solutions, although in those cases, it is always the same initial measurement ζ that is continually compared to each estimate, to track its convergence, while in this case we are allowing for that measurement to change over time.

2. *Compute the innovation covariance.* An important aspect of the Kalman filter, which we leverage, is that it considers the error covariances of the estimates and measurements. Let us denote the error covariance of the previous position estimate $\mathbf{P}_{k-1}$

$$\mathbf{P}_{k-1} = E\left\{\left(\mathbf{x}_{k-1} - \mathbf{x}\right)\left(\mathbf{x}_{k-1} - \mathbf{x}\right)^T\right\} \tag{7.7}$$

Given this value, the covariance of the innovation $\mathbf{y}_k$ is computed

$$\mathbf{S}_k = E\left\{\mathbf{y}_k\mathbf{y}_k^T\right\} = \mathbf{H}\left(\mathbf{x}_{k-1}\right)\mathbf{P}_{k-1}\mathbf{H}\left(\mathbf{x}_{k-1}\right)^T + \mathbf{C}_\zeta \tag{7.8}$$

where $\mathbf{H}(\mathbf{x})$ is the gradient of the measurement function, which represents how changes in the target's position translate to changes in the received measurements,[6] and $\mathbf{C}_\zeta$ is the measurement covariance matrix.

$$\mathbf{H}\left(\mathbf{x}_k\right) = \left.\frac{\partial \mathbf{z}(\mathbf{x})}{\partial \mathbf{x}}\right|_{\mathbf{x}=\mathbf{x}_k} = \mathbf{J}^T\left(\mathbf{x}_k\right) \tag{7.9}$$

Ideally we would supply the gradient with the true position $\mathbf{x}$, but that is not known, so the solution is to use the most recent estimated position $\mathbf{x}_{k-1}$. In this manner, our implementation follows the same approach as the *extended Kalman filter*.[7]

3. *Compute the Kalman gain.* The next step, the Kalman gain, is an attempt to optimally weight the two pieces of information (the prior position estimate and the innovation) to generate a new position estimate. The Kalman gain is given:

$$\mathbf{K}_k = \mathbf{P}_{k-1}\mathbf{H}^T\left(\mathbf{x}_{k-1}\right)\mathbf{S}_k^{-1} \tag{7.10}$$

4. *Update estimate.* Finally, we update the position estimate and its associated error covariance using the derived Kalman gain.

$$\mathbf{x}_k = \mathbf{x}_{k-1} + \mathbf{K}_k\mathbf{y}_k \tag{7.11}$$

$$\mathbf{P}_k = \left(\mathbf{I} - \mathbf{K}_k\mathbf{H}\left(\mathbf{x}_{k-1}\right)\right)\mathbf{P}_{k-1} \tag{7.12}$$

This algorithm is recursive, in that each update depends only on the new measurement ζ_k and the previous estimate $\mathbf{x}_{k-1}$ (as well as their respective covariance matrices $\mathbf{C}_\zeta$ and $\mathbf{P}_{k-1}$). To initialize it, we suggest using a standard geolocation solver for $\mathbf{x}_0$ from ζ_0 and using the CRLB at $\mathbf{x}_0$ as $\mathbf{P}_0$.

6. This is, conveniently, the transpose of the Jacobian $\mathbf{J}(\mathbf{x})$ that we have used extensively in this text.
7. The *extended Kalman filter* (EKF) is an extension of the Kalman filter (KF) that considers a nonlinear measurement function and a nonlinear transition function. While the KF is known to be optimal under Gaussian conditions, the EKF has no such proof. In place of the linear measurement and transition matrices, the EKF uses a first-order Taylor series approximation to linearize the system model [12]–[14].

These steps are implemented in the `tracker` package, as the function `ekfUpdate`. It is discussed in greater detail in the next chapter, but its usage is simple enough to explain here:

```
>[xk, Pk] = tracker.ekfUpdate(x_prev, P_prev, zeta, C, z, H);
```

where `x_prev` is $\widetilde{\mathbf{x}}_{k-1}$, `P_prev` is $\mathbf{P}_{k-1}$, `zeta` is ζ, `C` is $\mathbf{C}_{\zeta}$, `z` is a function handle for the measurement function $\mathbf{z}(\mathbf{x})$, and `H` is a function handle that computes that transpose of the Jacobian $\mathbf{H}(\mathbf{x}) = \mathbf{J}^T(\mathbf{x})$.

Example 7.3 Recursive Estimation of Two-Sensor DF

Consider a two-sensor AOA system at $(-5, 0)$ km and $(5, 0)$ km, detecting emissions from a Link-16 radio at (3, 25) km. The two sensors provide an AOA measurement every 1 ms, let us assume the accuracy of those measurements is $\sigma_\theta = 10°$.

Generate a series of noisy measurements corresponding to a 30-second observation interval. Use the EKF update algorithm to update an estimate of the target position with each measurement. Plot the trajectory of these estimates over time, as well as the distance between each update and the true position.

How long does it take for the position error to be less than 100 meters?

Solution: After defining the sensor positions (`x_aoa`), target position (`x_tgt`), and measurement covariance matrix (`C_df`), we compute the number of pulses required for the prescribed 30 s observation interval and generate noisy measurements.

```
pri = 1e-3;
T = 30; % observation period
num_pulses = floor(T/pri)+1;

psi = triang.measurement(x_aoa, x_tgt);
L = chol(C_df,'lower');
zeta = psi + L*randn(size(psi,1),num_pulses);
```

Next, we need to define function handles that will generate the measurement ($\mathbf{z}(\mathbf{x})$) and the gradient of the measurement function $\mathbf{H}(\mathbf{z})$ (recalling that the latter is simply the transpose of $\mathbf{J}(\mathbf{x})$):

```
z_fun = @(x) triang.measurement(x_aoa, x);
H_fun = @(x) triang.jacobian(x_aoa, x)';
```

Finally, after defining an initial estimate of (1, 1) km to seed the LS solver on the first iteration, we step over the pulses and call `ekfUpdate` with each measurement to update the estimated target position (`this_x`) and error covariance matrix (`this_P`).

```
for idx=1:num_pulses
    this_zeta = zeta(:,idx);
```

```
    if idx==1
        % Initialization
        this_x = triang.lsSoln(x_aoa, this_zeta, C_df, x_init);
        this_P = triang.computeCRLB(x_aoa, this_x, C_df);
    else
        % EKF Update
        [this_x, this_P] = tracker.ekfUpdate(prev_x, prev_P, this_zeta, C_df
            , z_fun, H_fun);
    end

    % Store the results and update the variables
    x_est(:, idx) = this_x;
    cep(idx) = utils.computeCEP50(this_P);
    prev_x = this_x;
    prev_P = this_P;
end
```

The trajectory of the solutions, along with the 1σ error ellipse for both a single measurement and the full set of observations, are plotted in Figure 7.5(a). To generate the ellipses, we used the CRLB using a single measurement (for the former), and the final value of `this_P` at the end of the simulation. In addition to this physical plot, we show the error, as a function of observation time, in Figure 7.5(b), along with the CEP_{50} computed from `this_P` after each update.

Note that the predicted error is not a steadily decreasing curve, as it would be if we simply scaled the error covariance, as we do in the sample mean solvers used earlier in this chapter. At each update, the innovation $\mathbf{y}_k$ provides information to the tracker on how well the previous estimate matches the newly received measurement. In the beginning, when there are a limited number of samples, each one generates a large innovation. It takes some time for the tracker to home in on a good solution and to start measuring residuals $\mathbf{y}_k$ that are in line with the provided measurement covariance $\mathbf{C}_\zeta$, indicating that the estimate $\mathbf{x}_k$ has converged to a stable solution. When this occurs, the reported error covariance $\mathbf{P}_k$ drops rapidly, as do the actual errors measured in our simulation.

To compute how long it takes for the error to be less than 100 m, we use the CEP_{50}, computed from the reported error covariance $\mathbf{P}_k$ after each iteration of `ekfUpdate`. Using this metric, we compute that 14.48 s of observations are required to achieve accuracy better than 100 m.

As before, we note that 14.48 s is a long time to observe a signal, given the realities of moving targets and things like communication frame duration and radar coherent processing intervals.

```
desired_cep = 100;
first_good_sample = find(cep < desired_cep,1,'first');
time_min = time_vec(first_good_sample);
```

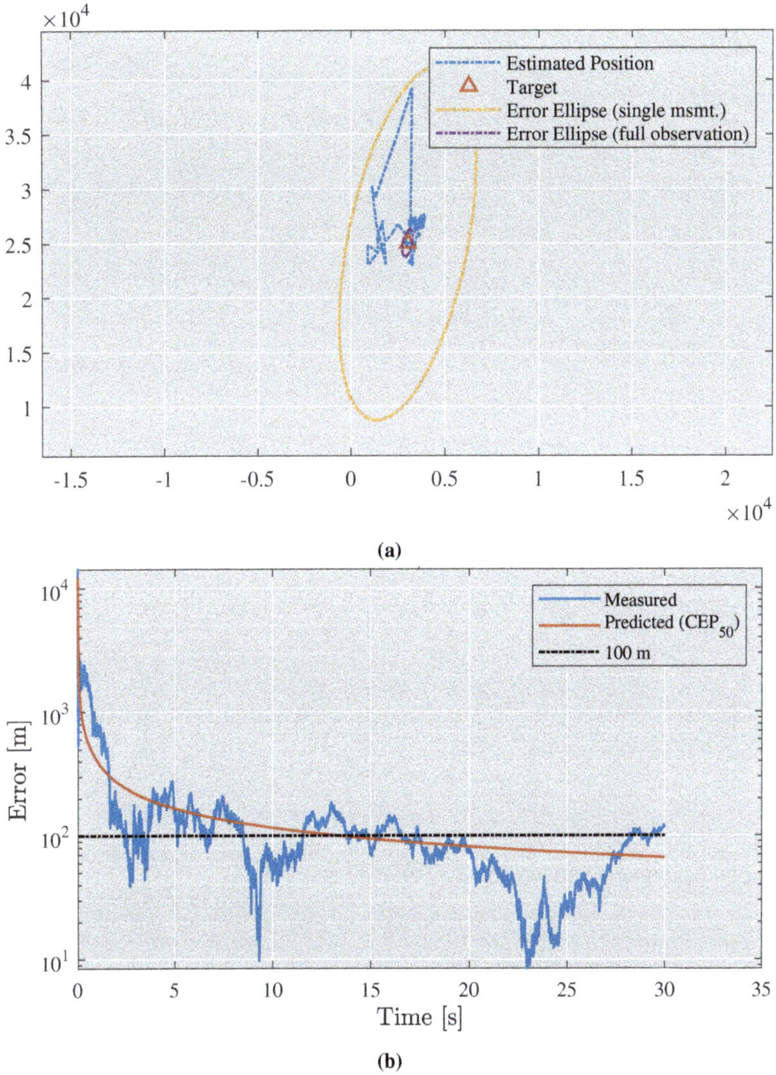

Figure 7.5 Geolocation results for the recursive two-sensor AOA scenario in Example 7.3, using the EKF-based solver. (a) Layout of the target area, error ellipse, and LS solution, and (b) measured and predicted geolocation error as a function of observation time.

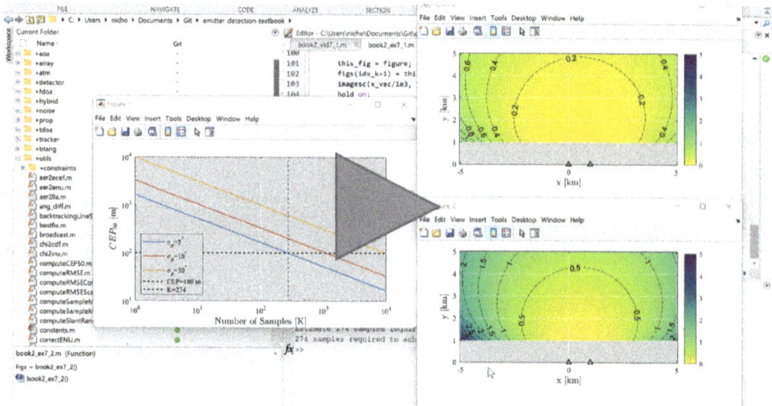

Figure 7.6 Video discussing the convergence of sample measurements in Example 7.5.

□

Figure 7.6 contains a link to a video discussion of Example 7.5. In this video, we discuss in a little more detail the convergence of error and the use of the sample mean to improve sensor performance.

7.2 GEOLOCATION OF A STATIONARY TARGET FROM MOVING SENSORS

The problem becomes slightly more complicated if either the target or the sensors are in motion. We'll address moving targets in the next chapter, but here we introduce the complexity of a moving sensor.

In the previous section, we discussed two methods of considering repeated measurements: computing the sample mean, and performing iterative geolocation. In this section, we'll discuss how they can be modified to work with moving sensors.

7.2.1 Measurement Space Solutions

Computing the sample mean no longer makes sense, as the measurements ζ_k are different at each time step k (as the sensors move, so too will their measurements change). We could, however, attempt to *track* this measurement as it changes, if the change in measurements is relatively slow compared to the rate at which they are taken.

Take, for example, the case shown in Figure 7.7(a), which depicts three TDOA sensors moving in formation, in the +y direction, with a single transmitter in their path.

The measurements generated by these sensors are heavily time-dependent, as shown in Figure 7.7(b).

There are many formal approaches, but in this case, a simple one is best for illustration. Consider the moving average of length N

$$\overline{\zeta}_k = \frac{1}{N}\sum_{n=0}^{N-1}\zeta_{k-n} \tag{7.13}$$

This is similar to the sample mean, except that we have limited the impact of older samples. If the measurements are approximately stationary, over the time interval N, it should function well.

A similar approach is exponential decay, which is computed:

$$\overline{\zeta}_k = \alpha\overline{\zeta}_{k-1} + (1-\alpha)\,\zeta_k,\ \alpha \in [0,1] \tag{7.14}$$

This is an example of an *infinite-impulse response* (IIR) filter, since the information from an earlier sample is never fully discarded, but its impact quickly grows negligible (particularly if $\alpha < .5$). By modifying α, one can control how much memory the filter has. Very small values of α mean it has little memory and behaves similarly to if we used only ζ_k to estimate position, where large values with $\alpha \approx 1$ result in a very long memory and behave similarly to a sample mean with large N.

The limitation, however, of working in measurement space is that we can't take advantage of the fact that sensor motion is known, and that we can't integrate over large changes in sensor position, as we're limited to windows over which ζ_k is approximately stationary. Thus, we will focus our attention on state space solutions.

7.2.2 State Space Solutions

If we operate in state space, then we can use the Kalman-inspired approach described in Section 7.1.2.1, since the parameter we are ultimately estimating ($\mathbf{x}$) is stationary, even over long time intervals and large changes in sensor position. The one modification to make is that the measurement function $\mathbf{z}(\mathbf{x})$ is now time-dependent due to sensor motion, as is its gradient $\mathbf{H}(\mathbf{x})$. We introduce the shorthand $\mathbf{z}_k(\mathbf{x})$ and $\mathbf{H}_k(\mathbf{x})$ to denote the time-varying measurement and Jacobian functions at time instant k. The Kalman update equations are now written:

$$\mathbf{y}_k = \zeta_k - \mathbf{z}_k(\mathbf{x}_{k-1}) \tag{7.15}$$
$$\mathbf{S}_k = \mathbf{H}_k(\mathbf{x}_{k-1})\,\mathbf{P}_{k-1}\mathbf{H}_k^T(\mathbf{x}_{k-1}) + \mathbf{C}_\zeta \tag{7.16}$$
$$\mathbf{K}_k = \mathbf{P}_{k-1}\mathbf{H}_k^T(\mathbf{x}_{k-1})\,\mathbf{S}_k^{-1} \tag{7.17}$$
$$\widehat{\mathbf{x}}_k = \mathbf{x}_{k-1} + \mathbf{K}_k\mathbf{y}_k \tag{7.18}$$
$$\mathbf{P}_k = (\mathbf{I} - \mathbf{K}_k\mathbf{H}_k(\mathbf{x}_{k-1}))\,\mathbf{P}_{k-1} \tag{7.19}$$

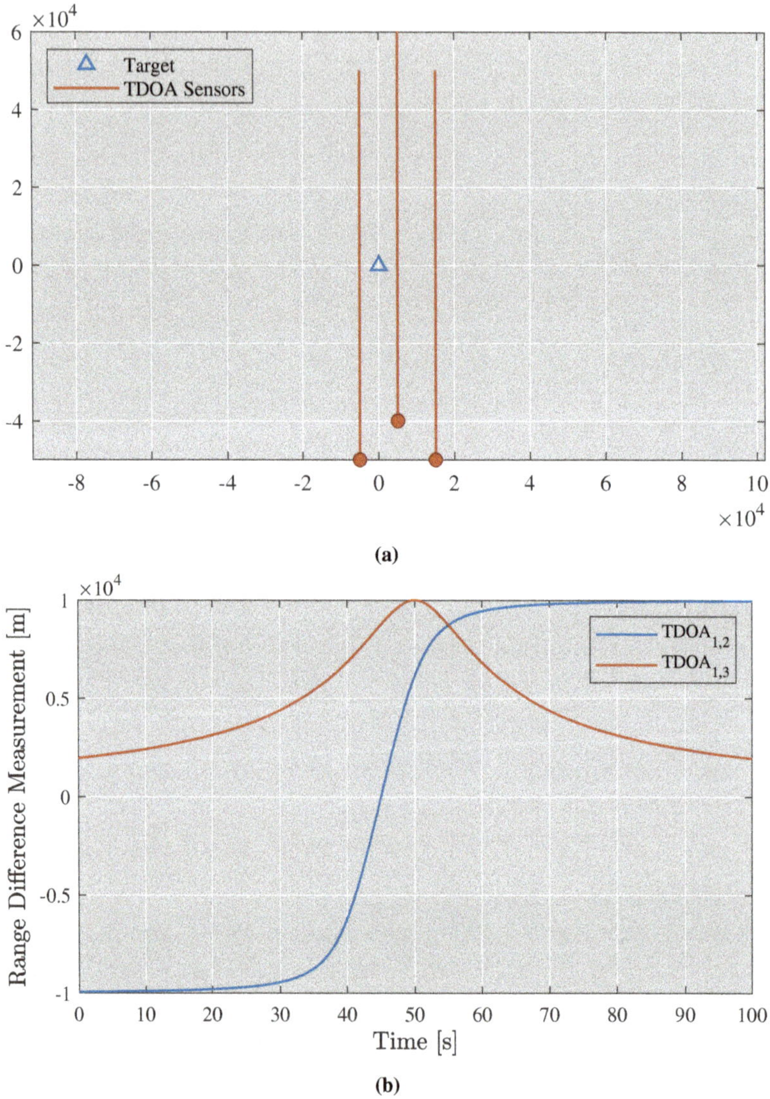

Figure 7.7 Example of a moving sensor TDOA scenario. (a) Layout and trajectory of the moving sensors and target, and (b) time-dependent range difference measurements.

To execute this recursive solver, we use the same function as before (`tracker.ekfUpdate`), but we feed it with a new function handle for both `z` and `H` on each iteration.

Example 7.4 Recursive AOA from Moving Sensors

Consider the geometry of Example 7.3, except that now the sensors have velocities of 200 m/s in the +y direction. Generate position estimates using data from a 30-second observation interval. Compare to the results from Example 7.3.

Solution: The setup and execution of this example is very similar to Example 7.3, with the exception that the sensors now move (we store their velocities in `v_aoa`) and that the measurement and measurement gradient functions are now time-dependent. The new loop over pulses is slightly more complex, including an update of the sensor positions, generation of a noisy measurement, and redefinition of the function handles.

```
for idx=1:num_pulses
    % Update positions
    this_x_aoa = x_aoa + v_aoa * (idx-1) * pri;

    % Update function handles
    z_fun = @(x) triang.measurement(this_x_aoa, x);
    H_fun = @(x) triang.jacobian(this_x_aoa, x)';

    % Generate noisy measurements
    psi = z_fun(x_tgt);
    zeta = psi + L*randn(num_dims,1);

    if idx==1
        % Initialization
        this_x = triang.lsSoln(this_x_aoa, zeta, C_df, x_init);
        this_P = triang.computeCRLB(this_x_aoa, this_x, C_df);
    else
        % EKF Update
        [this_x, this_P] = tracker.ekfUpdate(prev_x, prev_P, zeta, C_df,
            z_fun, H_fun);
    end

    % Store the results and update the variables
    x_est(:, idx) = this_x;
    cep(idx) = utils.computeCEP50(this_P);

    prev_x = this_x;
    prev_P = this_P;
end
```

Analogous plots to Figure 7.5(a) and (b) are shown in Figure 7.8(a) and (b), respectively. The principal difference is that we see a slight improvement in the convergence of geolocation errors. This is because we're relying not only on the additional measurements to improve sensor accuracy, but the changing baseline is giving us *geometric diversity*. The manifestation of that in this example is that the time required for a good (<100 m) geolocation estimate is reduced from 14.48 seconds to 11.05 seconds. □

Figure 7.9 contains a link to a video discussion of Example 7.4. In this video, we discuss in a little more detail the convergence of error and the use of the sample mean to improve sensor performance.

7.2.3 Complex Scenarios

Specification of the measurement function and gradient at each update affords us great flexibility. Not only can the measurement function specify changing sensor positions, but any other knowable change in parameters can be included, such as changes in sensor performance (via measurement error or biases), or the composition of sensors providing measurements.

For example, if measurements are arriving asynchronously from various sensors, then each track update can consider only the available measurements, by including different elements of $\mathbf{z}_k(\mathbf{x})$ at each time instance. Similarly, if measurements are missed, such as when a sensor is subject to interference, then those measurements can be easily excluded from ζ until the interference is suppressed.

For satellite-based scenarios, particularly those with satellites in low Earth orbit, any given sensor may only have line of sight to an emitter for a matter of minutes, and continuous coverage of an area of regard could require dozens, if not hundreds of satellites. At each time instance, only those satellites that have line of sight to an emitter need to be included in $\mathbf{z}_k(\mathbf{x})$.

These temporal variations are much more difficult to consider when computing the sample mean, which illustrates a distinct advantage of state-space solutions.

7.2.4 Degenerate Geometries

Due to this flexibility, state space solutions are robust to degenerate geometries, as they can use information from multiple time instances together to form a 3-D position, even if none of the individual measurements have sufficient information [15]. At each update, the available information is used to reshape the error covariance and predicted position, and any dimensions for which there is no information are simply *coasted* forward to the next update.

For example, consider Figure 7.10, which shows a single ground-based emitter (stationary), and the trajectory for an airborne AOA receiver in Figure 7.10(a). This aircraft is unable to generate a position estimate by itself from any one measurement. At the beginning of its trajectory, the receiver detects an incoming threat signal and solves

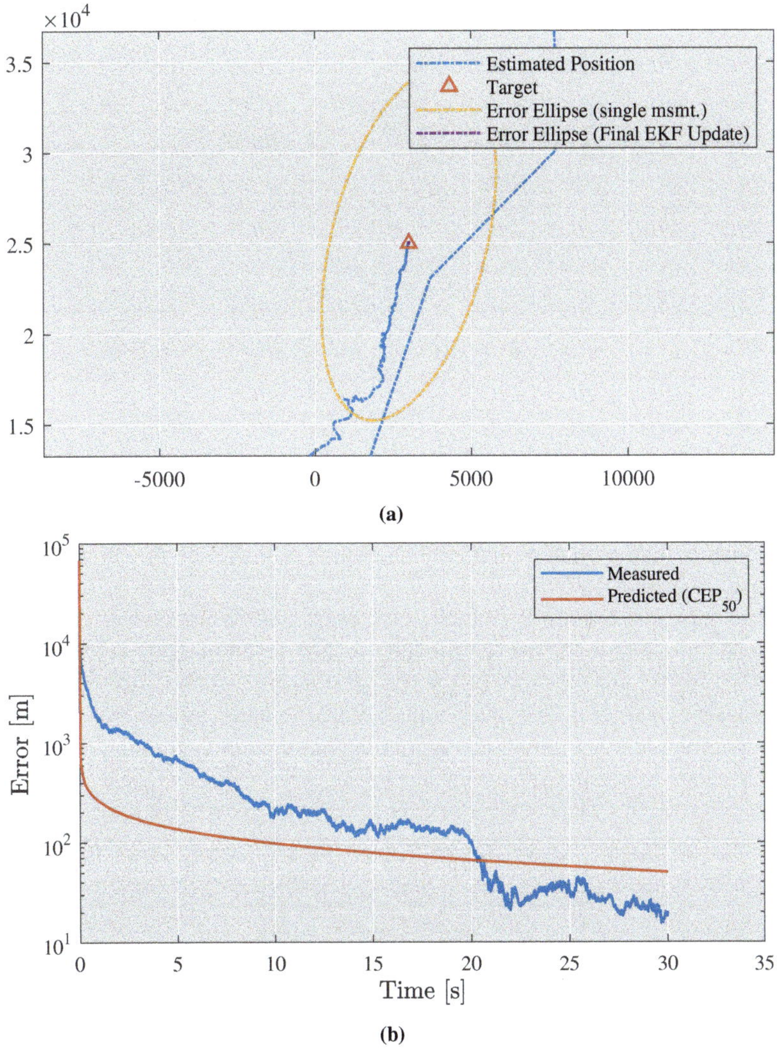

Figure 7.8 Geolocation results for the moving two-sensor AOA scenario in Example 7.4. (a) Layout of fixed target position, error ellipse, and estimated position (over time), and (b) error as a function of time, and the predicted CEP_{50} (computed from $\mathbf{P}_k$).

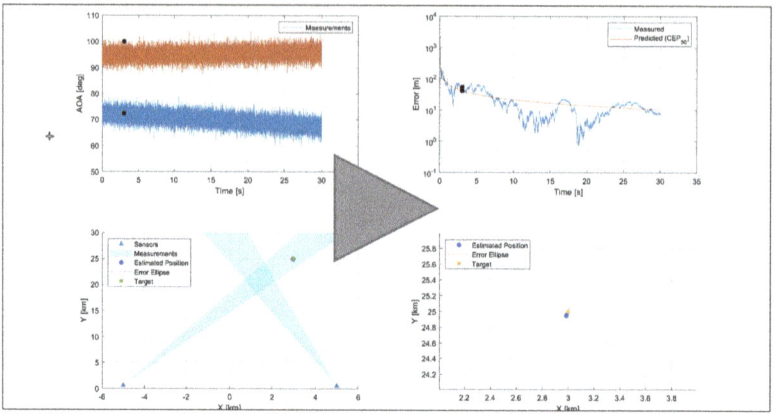

Figure 7.9 Video discussing the convergence of sample measurements in Example 7.4.

for its direction of arrival. This results in a line of bearing (LOB), rather than a pinpointed solution. From this LOB, however, the aircraft performs a maneuver designed to provide a baseline for additional measurements. Over time, the information from these additional measurements can be used to pinpoint the emitter, as shown by the progression of CEP over time in Figure 7.10(b).

The principal caveat is that the track must still be initialized, even if the first measurement is not sufficient to generate a 3-D position. The two most common approaches to this are (a) to collect multiple measurements and use them to form an initial position estimate in 3-D space, and (b) make an initial guess with very large initial error covariance. The former case is more robust, particularly for multitarget scenarios, where you want to be careful not to set the initial position error so large as to incorporate measurements from other sources. In Figure 7.10, we have chosen the latter approach for simplicity and made an initial guess of $1 - km$ for the y-coordinate with an error covariance of

$$P_0 = \begin{bmatrix} 1e6 & 0 \\ 0 & 100e6 \end{bmatrix} \tag{7.20}$$

This technique works for mobile emitters, so long as the receiver has greater maneuverability than the transmitter [16]. If that situation is reversed, then the problem is ill-posed, and infinite solutions will be possible. It is worth noting that significant research has been devoted to this challenging problem of angle-only tracking of moving targets, including the proposal of many advanced trackers using formulations more complex than the Kalman filter that we have used as our baseline for discussion **10.1016/S0165-1684(03)00042-2**, [17]–[21].

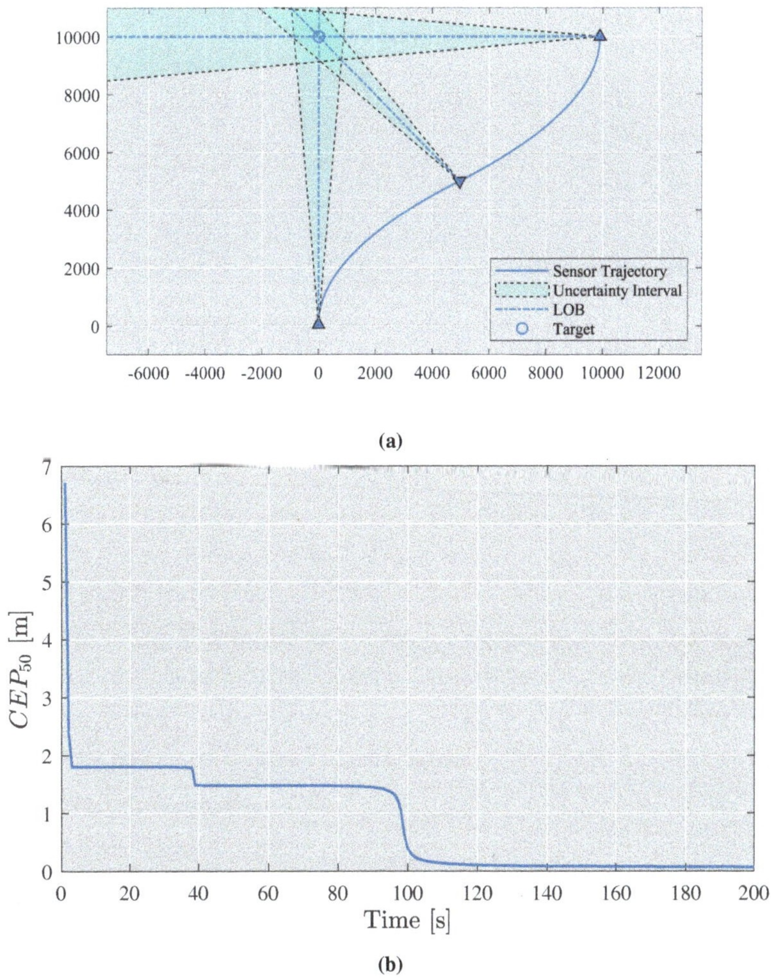

Figure 7.10 Example of a degenerate geolocation scenario with a single airborne sensor performing AOA measurements against a stationary ground emitter. (a) Layout of the scenario, target, aircraft trajectory, and three example AOA measurements with uncertainty, and (b) error (CEP_{50} computed from the EKF-inspired state space solver) as a function of time.

7.3 CONCLUSION

In this chapter, we discussed how repeated measurements can improve geolocation performance. We introduced measurement space and state space solutions and showed how moving sensors can provide additional information when fed into something like the extended Kalman filter.

7.4 PROBLEM SET

7.1 A hybrid geolocation system is said to be capable of achieving an RMSE of 1 km for a given target signal. How many *independent* signal samples must the system collect in order to achieve 100 m error? 25 m?

7.2 A TDOA system is designed to detect Chinese 3G signals using the *time-division synchronous code-division multiple access* (TD-SCDMA) signal, for which 5-ms *frames* are divided into 7 time slots, of 675 μs each. Assume for simplicity that the target is allocated one time slot in each frame and that a single TDOA measurement involves listening to the entire time slot (to gain sufficient SNR for time difference estimation). If the CRLB for a single measurement is 5 km, what is the position error after 5 seconds? Assume, for simplicity, that the transmitter and receiver are stationary.

7.3 Repeat Example 7.2 using gradient descent. Comment on the differences.

7.4 Repeat Example 7.3. Compute a recursive sample mean, using (7.14) with a decay parameter of $\alpha = .7$. Use the least square solver to reestimate the target's position with each new sample. Repeat with $\alpha = .3$. Comment on the differences.

7.5 Repeat Example 7.4, but with two sensors in formation, a second 500 meters to the right of the first, conducting TDOA instead of DF, with a timing accuracy of 1 μs.

References

[1] L. L. Scharf, *Statistical Signal Processing*. Reading, MA: Addison-Wesley, 1991.

[2] N. A. O'Donoughue, *Emitter Detection and Geolocation for Electronic Warfare*. Norwood, MA: Artech House, 2019.

[3] C.-H. Kao, "Performance analysis of a JTIDS/Link-16-type waveform transmitted over slow, flat Nakagami fading channels in the presence of narrowband interference," Ph.D. dissertation, Naval Postgraduate School, 2008.

[4] M. A. Richards, *Fundamentals of Radar Signal Processing*. New York, NY: McGraw-Hill Education, 2014.

[5] R. E. Kalman, "A new approach to linear filtering and prediction problems," *Transactions of the ASME–Journal of Basic Engineering*, 1960.

[6] R. Faragher, "Understanding the basis of the Kalman filter via a simple and intuitive derivation [lecture notes]," *IEEE Signal Processing Magazine*, vol. 29, no. 5, pp. 128–132, 2012. DOI: 10.1109/MSP.2012.2203621.

[7] P. Hargrave, "A tutorial introduction to Kalman filtering," *IEE Colloquium on Kalman Filters: Introduction, Applications and Future Developments*, pp. 1/1–1/6, 1989.

[8] D. P. Mandic, S. Kanna, and A. G. Constantinides, "On the intrinsic relationship between the least mean square and Kalman filters [lecture notes]," *IEEE Signal Processing Magazine*, vol. 32, no. 6, pp. 117–122, 2015. DOI: 10.1109/MSP.2015.2461733.

[9] J. Humpherys and J. West, "Kalman filtering with Newton's method [lecture notes]," *IEEE Control Systems Magazine*, vol. 30, no. 6, pp. 101–106, 2010. DOI: 10.1109/MCS.2010.938485.

[10] A. Kheirati Roonizi, "An efficient algorithm for maneuvering target tracking [tips amp; tricks]," *IEEE Signal Processing Magazine*, vol. 38, no. 1, pp. 122–130, 2021. DOI: 10.1109/MSP.2020.3029386.

[11] M. S. Grewal and A. P. Andrews, *Kalman Filtering: Theory and Practice with MATLAB*. John Wiley & Sons, 2014.

[12] F. Daum, "Nonlinear filters: Beyond the Kalman filter," *IEEE Aerospace and Electronic Systems Magazine*, vol. 20, no. 8, pp. 57–69, 2005. DOI: 10.1109/MAES.2005.1499276.

[13] K. Spingarn, "Passive position location estimation using the extended Kalman filter," *IEEE Transactions on Aerospace and Electronic Systems*, vol. AES-23, no. 4, pp. 558–567, 1987. DOI: 10.1109/TAES.1987.310889.

[14] S. Fu, J. Vian, and D. Grose, "Determination of ground emitter location," *IEEE Aerospace and Electronic Systems Magazine*, vol. 3, no. 12, pp. 15–18, 1988. DOI: 10.1109/62.9376.

[15] M. Skolnik, *Radar Handbook, 3rd Edition*. New York, NY: McGraw-Hill Education, 2008.

[16] A. Logothetis, A. Isaksson, and R. Evans, "Comparison of suboptimal strategies for optimal own-ship maneuvers in bearings-only tracking," *Proc. of the 1998 American Control Conference. ACC (IEEE Cat. No.98CH36207)*, vol. 6, pp. 3334–3338, 1998. DOI: 10.1109/ACC.1998.703192.

[17] R. Karlsson and F. Gustafsson, "Range estimation using angle-only target tracking with particle filters," *Proc. of the 2001 American Control Conference. (Cat. No.01CH37148)*, vol. 5, pp. 3743–3748, 2001. DOI: 10.1109/ACC.2001.946218.

[18] A. Holtsberg and J. Holst, "Estimation and confidence in bearings only tracking," *Proc. of the Twenty-Fifth Asilomar Conference on Signals, Systems & Computers (Asilomar '91)*, vol. 2, pp. 883–887, 1991. DOI: 10.1109/ACSSC.1991.186573.

[19] V. Aidala and S. Hammel, "Utilization of modified polar coordinates for bearings-only tracking," *IEEE Transactions on Automatic Control*, vol. 28, no. 3, pp. 283–294, 1983. DOI: 10.1109/TAC.1983.1103230.

[20] S. Arulampalam and B. Ristic, "Comparison of the particle filter with range-parameterized and modified polar ekfs for angle-only tracking," *Signal and Data Processing of Small Targets 2000*, vol. 4048, pp. 288–299, 2000.

[21] R. Allen and S. Blackman, "Implementation of an angle-only tracking filter," in *Defense, Security, and Sensing*, 1991.

Chapter 8

Geolocation of Moving Targets

8.1 INTRODUCTION

In this chapter, we discuss geolocation of moving targets from their emissions. We will use the concept of a tracker as the basis of our discussion, specifically a Bayesian approach that consists of a two-part *predict-update* cycle, as exemplified by the Kalman filter [1], [2].

The KF has been widely employed and studied since its introduction in 1960 [1]. It enjoys wide adoption and robust performance (when properly initialized), but has a critical limitation: it is a *linear* system. The KF can only be applied directly if the dynamics of a target's state and the relationship between that state and the observable measurements are both linear. Since the measurement functions that we use are nonlinear, we will focus on an extension known as the *extended Kalman filter* (EKF)[3].

There are many other tracking approaches that are relevant to moving target passive geolocation, which we briefly discuss in Section 8.4. For brevity, we focus on the EKF in this text and treat it as the baseline against which more novel techniques can be compared.

8.2 AN INTRODUCTION TO TRACKING

As mentioned before, we will leverage a two-step predict-update cycle, shown in Figure 8.1. Consider a series of discrete updates, indexed by k. At each time instance, a measurement ζ_k is produced, distributed according to:

$$\zeta_k \sim \mathcal{N}\left(\mathbf{z}_k(\mathbf{x}), \mathbf{C}_\zeta\right) \tag{8.1}$$

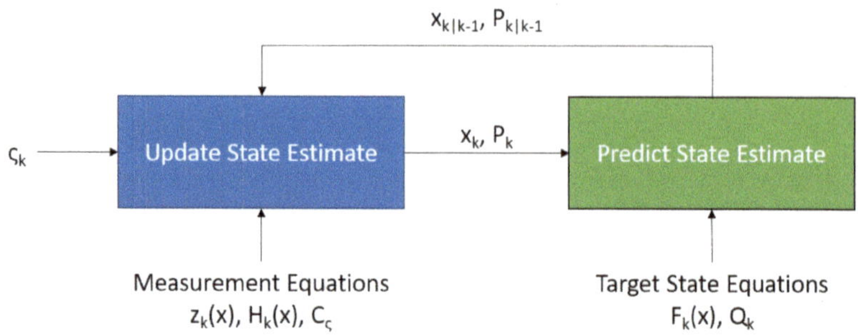

Figure 8.1 Illustration of the predict-update cycle of a tracker loop.

for some measurement function $\mathbf{z}_k(\mathbf{x})$ and is related to the target's state $\mathbf{x}$ with the function.[1]

$$\mathbf{H}_k(\mathbf{x}) = \frac{\partial \mathbf{z}_k(\mathbf{x})}{\partial \mathbf{x}} = \mathbf{J}^T(\mathbf{x}) \tag{8.2}$$

In addition to these measurements, we also have the predicted target state at time k, given information available up until time $k-1$, denoted $\widehat{\mathbf{x}}_{k|k-1}$ and its variance $\mathbf{P}_{k|k-1}$.

The task now is to first *update* this prediction given the new measurement ζ_k, yielding the estimate $\widehat{\mathbf{x}}_{k|k}$ and its corresponding variance $\mathbf{P}_{k|k}$, then to predict forward to the next time instant, generating the prediction $\widehat{\mathbf{x}}_{k|+1|k}$ and corresponding variance $\mathbf{P}_{k+1|k}$.

We will present a brief summary of the KF and EKF algorithms. This is far from a detailed derivation, and is not intended to demonstrate their optimality or to show every stage of how they were derived. Interested readers are referred to one of the countless tutorials or texts on the subject, such as [4]–[11].

8.2.1 Update Stage

For the update stage, we seek to estimate $\widehat{\mathbf{x}}_{k|k}$, given the prediction $\widehat{\mathbf{x}}_{k|k-1}$ and the measurement ζ_k. To do this, we write the measurement model

$$\zeta_k = \mathbf{z}_k(\mathbf{x}) + \mathbf{v}_k \tag{8.3}$$

where $\mathbf{v}_k$ is the measurement noise, with covariance matrix $\mathbf{C}_\zeta$.

In order to use the KF framework, we replace the nonlinear function $\mathbf{z}(\mathbf{x})$ with its Taylor series representation, using a first-order Taylor series centered at the most recent

1. Note here that $\mathbf{x}$ includes, but is not limited to, the position and velocity of the target in each dimension. This is a slightly expanded definition of the term relative to how it has been used in the first seven chapters of this text.

predicted position; the result is referred to as the EKF:[2]

$$\mathbf{z}(\mathbf{x}) \approx \mathbf{z}\left(\widehat{\mathbf{x}}_{k|k-1}\right) + \mathbf{H}_k\left(\widehat{\mathbf{x}}_{k|k-1}\right)\left(\mathbf{x} - \widehat{\mathbf{x}}_{k|k-1}\right) \tag{8.4}$$

With this approximation, we can rewrite the measurement function

$$\zeta_k \approx \mathbf{z}\left(\widehat{\mathbf{x}}_{k|k-1}\right) + \mathbf{H}_k\left(\widehat{\mathbf{x}}_{k|k-1}\right)\left(\mathbf{x} - \widehat{\mathbf{x}}_{k|k-1}\right) + \mathbf{v}_k \tag{8.5}$$

To simplify, we consider the innovation (or residual), which is the difference between the received and expected measurements. We have used the term $\mathbf{y}$ for similar difference measurements in the derivation of iterative least squares and gradient descent algorithms, in [12], and do so again here.

$$\begin{aligned} \mathbf{y}_k &= \zeta_k - \mathbf{z}_k\left(\widehat{\mathbf{x}}_{k|k-1}\right) & (8.6) \\ &\approx \mathbf{H}_k\left(\widehat{\mathbf{x}}_{k|k-1}\right)\left(\mathbf{x} - \widehat{\mathbf{x}}_{k|k-1}\right) + \mathbf{v}_k & (8.7) \end{aligned}$$

The error covariance of the innovation is computed

$$\begin{aligned} \mathbf{S}_k &= E\left\{\mathbf{y}_k\mathbf{y}_k^T\right\} & (8.8) \\ &= \mathbf{H}_k\left(\widehat{\mathbf{x}}_{k|k-1}\right)\mathbf{P}_{k|k-1}\mathbf{H}_k^T\left(\widehat{\mathbf{x}}_{k|k-1}\right) + \mathbf{C}_\zeta & (8.9) \end{aligned}$$

Next, we define the update step as a linear combination of the prediction $\left(\widehat{\mathbf{x}}_{k|k-1}\right)$ and the innovation $(\mathbf{y}_k)$.

$$\widehat{\mathbf{x}}_{k|k} = \widehat{\mathbf{x}}_{k|k-1} + \mathbf{K}_k\mathbf{y}_k \tag{8.10}$$

where $\mathbf{K}_k$ is a weight matrix used to update the state estimate, referred to as the *Kalman Gain*. The error covariance of the updated state estimate is computed:

$$\begin{aligned} \mathbf{P}_{k|k} =& E\left\{\left[\mathbf{x}_k - \widehat{\mathbf{x}}_{k|k}\right]\left[\mathbf{x}_k - \widehat{\mathbf{x}}_{k|k}\right]^T\right\} & (8.11) \\ =& \left(\mathbf{I} - \mathbf{K}_k\mathbf{H}_k\left(\widehat{\mathbf{x}}_{k|k-1}\right)\right)\mathbf{P}_{k|k-1}\left(\mathbf{I} - \mathbf{K}_k\mathbf{H}_k\left(\widehat{\mathbf{x}}_{k|k-1}\right)\right) & (8.12) \\ &+ \mathbf{K}_k\mathbf{C}_\zeta\mathbf{K}_k^T & (8.13) \end{aligned}$$

What remains is to solve for the ideal Kalman gain. Under the criteria of minimizing the trace of $\mathbf{P}_{k|k}$, it can be shown that:

$$\mathbf{K}_k = \mathbf{P}_{k|k-1}\mathbf{H}_k^T\left(\widehat{\mathbf{x}}_{k|k-1}\right)\mathbf{S}_k^{-1} \tag{8.14}$$

and by substituting (8.14) into (8.13), we can simplify the calculation of $\mathbf{P}_{k|k}$:

$$\mathbf{P}_{k|k} = \left(\mathbf{I} - \mathbf{K}_k\mathbf{H}_k\left(\widehat{\mathbf{x}}_{k|k-1}\right)\right)\mathbf{P}_{k|k-1} \tag{8.15}$$

Calculation of (8.10) and (8.15) is implemented in the function `ekfUpdate` in the `tracker` package. The linear version is implemented in `kfUpdate`. The usage is:

2. See Chapter 3 of [11] for a derivation.

```
>[xk, Pk] = tracker.ekfUpdate(x_prev, P_prev, zeta, C, z, H);
```

where x_prev is $\widehat{\mathbf{x}}_{k|k-1}$, P_prev is $\mathbf{P}_{k|k-1}$, zeta is ζ, C is $\mathbf{C}_\zeta$, z is a function handle for the measurement function $\mathbf{z}(\mathbf{x})$, and H is a function handle that computes that transpose of the Jacobian $\mathbf{H}(\mathbf{x}) = \mathbf{J}^T(\mathbf{x})$.

8.2.2 Prediction Stage

The next step is to predict forward to the next time step. In this section, we'll predict forward to time k using information up until time $k-1$, but the math is the same for predicting forward to time $k+1$ from information available at time k.

To do this, we represent the relationship between the target state at time $k-1$ and k with a state transition matrix

$$\mathbf{x}_k = \mathbf{F}\mathbf{x}_{k-1} + \mathbf{w}_k \tag{8.16}$$

where $\mathbf{w}_k$ is a zero-mean Gaussian random vector with covariance matrix $\mathbf{Q}$. The former represents the model of system dynamics, such as Newton's laws of kinematic motion, and is referred to as a *state transition matrix*, while the latter is *process noise*, which allows us to account for random perturbations in target state (such as turbulence, or the impact of unmodeled control inputs).

Proper selection of a state transition matrix $\mathbf{F}$ and process noise covariance matrix $\mathbf{Q}$ is critical to tracker performance. We will discuss system models briefly in the next section. Regardless of the model selected, however, the prediction process is straightforward

$$\widehat{\mathbf{x}}_{k|k-1} = \mathbf{F}\widehat{\mathbf{x}}_{k-1|k-1} \tag{8.17}$$

and the error covariance of the predicted state is

$$\mathbf{P}_{k|k-1} = \mathbf{F}\mathbf{P}_{k-1|k-1}\mathbf{F}^T + \mathbf{Q} \tag{8.18}$$

For a nonlinear model, we replace the state transition matrix $\mathbf{F}$ with a nonlinear function $\mathbf{f}(\mathbf{x})$. Similar to the update stage, we use a first-order Taylor series representation to approximate $\mathbf{f}(\mathbf{x})$ with a linear function, which requires the gradient at the most recent estimated position $\mathbf{F}_{k-1}$

$$\mathbf{F}_{k-1} = \left.\frac{\partial \mathbf{f}(\mathbf{x})}{\partial \mathbf{x}}\right|_{\mathbf{x}=\widehat{\mathbf{x}}_{k-1|k-1}} \tag{8.19}$$

With these definitions, the prediction and prediction error covariance become:

$$\widehat{\mathbf{x}}_{k|k-1} = \mathbf{f}\left(\widehat{\mathbf{x}}_{k-1|k-1}\right) \tag{8.20}$$

$$\mathbf{P}_{k|k-1} = \mathbf{F}_{k-1}\mathbf{P}_{k-1|k-1}\mathbf{F}_{k-1}^T + \mathbf{Q} \tag{8.21}$$

This is the EKF's form of the prediction stage. The linear model is implemented in the function `kfPredict` and the nonlinear model in `ekfPredict`, both in the `tracker` package. The former accepts the state transition matrix $\mathbf{F}$, while the latter accepts a function handle `F_fun` that accepts a position estimate $\widehat{\mathbf{x}}_{k|k}$ and returns the matrix $\mathbf{F}_k$.

A linear KF state transition model can be mixed with a nonlinear EKF measurement model, as we will do, and vice versa. There is no requirement to use all linear or all nonlinear models.

8.2.3 Limitations

The Kalman filter is optimal under a limited set of circumstances. Namely, that (1) the system and measurement models are linear, (2) the system and measurement models are known completely, (3) the random components (process noise and measurement noise) are Gaussian with known expectation and variance, and (4) the system is completely observable.[3] If those conditions are all met, then the trace of the covariance matrix at each update is minimized.

The EKF, by contrast, is not guaranteed to be optimal. Indeed, it has been shown not to achieve the CRLB in some cases relevant to our text. See, for example, [14], in which an EKF-based tracker for a ballistic target such as a satellite was compared to an iterative LS state estimator and found to underperform the CRLB (while the iterative LS estimator achieved it).

8.2.4 Coordinate Selection

Tracking can be accomplished in sensor coordinates (typically local spherical, AER), or in a global coordinate system (such as ECEF, LLA, or ECI, discussed in Chapter 4).

Typically, selection of a local coordinate system, particularly a local spherical one, makes the most sense in a single-sensor geometry. In this chapter, and the accompanying code, we will work in Cartesian coordinates. It does not matter if those coordinates are local or global, so long as all of the provided inputs (position and velocity of sensors and targets) are in the same reference frame.

There can be advantages to working in a spherical reference frame, particularly for tracking satellites, which have simplified motion models in orbiting reference frames like ECI, but for simplicity of implementation, we have restricted ourselves to Cartesian coordinates.

3. *Observability* is the notion that a system's state, or specific elements of its state, can be inferred at time k_1 from knowledge of the state at a previous time k_0 and the available observations z_k between them [13]. A system can be *partially observable*. For example, a target in 3-D space is only partially observable by TDOA sensors that all lie on a plane (such as when they are all coaltitude).

8.3 STATE SPACE MODELS

In this section, we will discuss the selection of appropriate state space models, namely the choice of a kinematic model to describe target motion from one time instance to the next (sometimes referred to as a *state transition* model) and a measurement model that describes the relationship between target states and the measurements generated.

8.3.1 State Space Representation of Targets

The state vector, which we denote as $\mathbf{x}$, is a combination of position, velocity, and any other relevant parameters that aid in tracking a target. We will discuss these in more detail in the next section, but any parameter that affects the kinematic model of target motion, or perhaps even affects how a target's features are measured, is a valid target state.

Most of the time, the target state will consist of position, velocity, and (occasionally) acceleration, in each of the spatial dimensions. There is no required order to how they are included in the state vector $\mathbf{x}$, but we have chosen to group them by parameter *type* rather than by axis.

In other words, our state vector representation is:

$$\mathbf{x} = \left[x,\ y,\ z,\ v_x,\ v_y,\ v_z,\ a_x,\ a_y,\ a_z \ldots\right]^T \tag{8.22}$$

where any additional parameters are listed after the acceleration terms.

The format for how $\mathbf{x}$ is defined in our models will be specified in a struct called `state_space` that will be returned when the kinematic model is instantiated and will need to be passed to some of the `tracker` functions, to ensure that they index the states appropriately. Users may wish to know that there is a field `pos_idx` that can be used to extract the spatial coordinates, in order to plot estimates, or to compute error statistics on position.

For example, if `x_est` has a collection of state estimates over time, and `P_est` has the corresponding estimation error covariance matrices, we can extract the position estimates and error with:

```
x_pos = x(state_space.pos_idx,:);
P_pos = P(state_space.pos_idx,state_space.pos_idx,:);
```

8.3.2 Kinematic Models

A distinct advantage of Cartesian coordinate tracking is that it affords several convenient moving target transition models. Let us assume the interval from one update to the next is T seconds. From Newton's equations of motion, we have a few simple motion models to choose from:

$$x_k = x_{k-1} + Tv_{k-1} \qquad \text{constant velocity} \tag{8.23}$$

$$x_k = x_{k-1} + Tv_{k-1} + \frac{1}{2}T^2 a_{k-1} \qquad \text{constant acceleration} \tag{8.24}$$

$$x_k = x_{k-1} + Tv_{k-1} + \frac{1}{2}T^2 a_{k-1} + \frac{1}{6}T^3 j_{k-1} \qquad \text{constant jerk} \tag{8.25}$$

where x_k is the position (in a single dimension) at time k, x_{k-1} is its position at the previous time step, v_{k-1} is its velocity, a_{k-1} is its acceleration, and j_{k-1} is its jerk in that dimension (derivative of acceleration). In general, we can continue to increase the order of our kinematic model indefinitely. As complexity increases, the tracker's ability to follow complex target motion increases as well. The downside, however, is that increasing the model order increases the number of parameters to track, which requires additional observations and comes with the risk of overfitting the model to the available data.

Using the kinematic models above, it is straightforward to construct the state transition matrix $\mathbf{F}$. For simplicity, we'll assume the CA model here, in which case the target's state has 3N components (position, velocity, and acceleration in N dimensions). The discrete-time system model, including random acceleration is:

$$\mathbf{F} = \begin{bmatrix} \mathbf{I}_N & T\mathbf{I}_N & \frac{T^2}{2}\mathbf{I}_N \\ \mathbf{0}_{NxN} & \mathbf{I}_N & T\mathbf{I}_N \\ \mathbf{0}_{NxN} & \mathbf{0}_{NxN} & \mathbf{I}_N \end{bmatrix} \tag{8.26}$$

$$\mathbf{w} \sim \mathcal{N}(\mathbf{0}, \mathbf{Q}) \tag{8.27}$$

$$\mathbf{Q} = \sigma_a^2 \begin{bmatrix} \frac{T^4}{4}\mathbf{I}_N & \frac{T^3}{2}\mathbf{I}_N & \frac{T^2}{2}\mathbf{I}_N \\ \frac{T^3}{2}\mathbf{I}_N & T^2\mathbf{I}_N & T\mathbf{I}_N \\ \frac{T^2}{2}\mathbf{I}_N & T\mathbf{I}_N & \mathbf{I}_N \end{bmatrix} \tag{8.28}$$

where σ_a^2 is the variance of random acceleration (assumed to be the same in all three dimensions), and N is the number of spatial dimensions modeled (2 or 3).

The specific equations in use vary greatly from model to model, and in many cases are nonlinear. Table 8.1 lists several well-known kinematic models. For simplicity, we implement only the first three (CA, CV, and CJ) in the provided MATLAB® code, and the rest are left as an exercise to the reader.

To define a kinematic model in the provided code, call the function `makeKinematicModel` of the `tracker` package. It will generate three outputs: (1) a function handle `f_fun` that accepts an update interval T (in seconds) and generates the transition matrix $\mathbf{F}$, (2) a function handle `Q_fun` that accepts both an update interval T and an acceleration covariance matrix `covar` that dictates the continuous time covariance of random acceleration terms in each dimension, and (3) a structure `state_space` that dictates how the state space $\mathbf{x}$ is constructed. The function is called with a string dictating which model to return.

Table 8.1 Some Common Kinematic Motion Models

Model	States	Common Usage	Citations
Constant velocity (CV)	Position, velocity	Level flight	[15], [16]
Constant acceleration (CA)	Position, velocity, acceleration	Level turns	[15], [16]
Constant jerk (CJ)	Position, velocity, acceleration, jerk	Highly agile maneuvering targets	[15], [17]
Ballistic trajectory	Position, velocity	Nonmaneuvering ballistic targets (e.g., satellites, mortars)	[15], [18]
Aerodynamically maneuvering targets	Position, velocity, ballistic coefficient†, lift coefficient†	Gliding aircraft and weapons	[15], [18]
Ballistic reentry vehicles	Position, velocity, ballistic coefficient†	Nonmaneuvering ballistic targets on reentry	
Maneuvering reentry vehicles (MaRV)	Position, velocity, ballistic coefficient†, turn coefficient†, and climb coefficient†	Ballistic reentry vehicles with the ability to maneuver	[15], [18], [19]

†Coefficients are required only once in the state vector, as opposed to position, velocity, and acceleration states, which are replicated for each spatial dimension in the coordinate system.

```
accel_covar = diag([1, 1, .1]); % 1 m/s^2 in x/y, .1 m/s^2 in z
[f_fun, q_fun, state_space] = tracker.makeKinematicModel('ca', num_dims,
    process_covar);

update_rate = 60; % seconds between tracker updates
F = f_fun(update_rate);
Q = q_fun(update_rate, accel_covar);
```

We will discuss in the next section how to set the acceleration covariance.

The choice of kinematic model is very important. For example, consider a nonmaneuvering aircraft. The CV model will perform the best, while a CA model will track the target, but will have increased error compared to the CV filter, since it is estimating acceleration (which is zero for a nonmaneuvering aircraft). The difference is akin to the improvement seen in Chapter 5 when altitude is constrained. If, however, the target maneuvers, then the CV model will be unable to track it without some modification (such as a subroutine to detect increased track uncertainty and increase process noise assumptions to accommodate the maneuver), while a CA model will inherently estimate and track the maneuver. This example illustrates that not only is it critical to select the right model, but that multiple models may be needed. This is the motivation behind higher-order track architectures, such as *interacting multiple models*, in which multiple trackers are run in parallel with different kinematic models, and at each time step the prediction that most closely matches the observed data is kept [20]–[24].

8.3.3 Process Noise

Closely linked to the target kinematic model is the choice of a process noise model; indeed most kinematic models explicitly define the appropriate process noise assumptions. For example, with both a constant velocity model, the model for process noise is random acceleration (zero mean), while for the constant acceleration model, it is a Markov process that also results in random acceleration process noise, albeit this time as perturbations from the tracked acceleration state.

With other models, such as ballistic flight, the choice can be more nuanced. For example, a ballistic target may have different motion models in its three phases of flight (boost, coast, and reentry), each of which with its own process noise assumptions. During the boost phase, process noise is likely dominated by perturbations in the thrust generated (random acceleration primarily in the direction of travel), while during coast it is much smaller and dominated by drag (again, primarily in the direction of travel, but with a more noticeable component perpendicular to the direction of travel, as drag can cause rotation or lateral forces if the target is not perfectly stabilized, and fluctuations in the Earth's gravitational field will affect acceleration in the direction of the Earth), and during reentry it will likely be dominated by turbulent airflow, imperfections, and asymmetries in the reentry vehicle and any maneuvers that are made (if they are not explicitly tracked) [15].

Each of these process noise models comes with one or more parameters that must be set, typically the variance on the random acceleration term at each time index. Selection of that parameter is difficult and has tremendous impact on filter performance. Some methods have been discussed for automatic selection of the parameters (see, for example, [25]), and research is ongoing. In many cases, the selection boils down to a tuning process, wherein representative targets are modeled, and tracks are simulated with varying process noise parameters.

8.3.4 Measurement Models

For measurement models, the process is very straightforward, particularly given the work we have already done to define measurement functions for each of the passive geolocation sensor phenomenologies. The nonlinear measurement function $\mathbf{z}(\mathbf{x})$ is already defined in the `measurement` function of the `triang`, `tdoa`, `fdoa`, and `hybrid` packages. The only difference is that now the input $\mathbf{x}$ is the full state of the target, not merely its position. To handle this, we can make an anonymous function that maps from state to position vector. For example, to modify the `measurement` function of an AOA scenario, we simply write two lines.

```
x_pos = @(x) x(state_space.pos_idx,:);
z_fun = @(x) triang.measurement(x_aoa, x_pos(x));
```

The first function excises the position coordinates from the larger state vector, using the `state_space` struct that was returned when the kinematic model was defined. The second line uses that converter to create an anonymous function `z_fun` that will accept the state and return AOA measurements. This should be done within a tracker loop if the sensors are moving, so that the position `x_aoa` can be updated on each iteration. A similar extraction is necessary for target velocity when modeling FDOA measurements.

Similarly, the gradient $\mathbf{H}(\mathbf{x})$ is simply the transpose of the Jacobian returned by the `jacobian` function of each package, with a slight modification. Just as with the measurement, the Jacobian needs to consider only the position entries (or position and velocity if it is an FDOA scenario). In addition, however, the Jacobian needs to be modified to include empty columns, to indicate that certain states have no impact on the measurement. For an illustration of this, see Figure 8.2. Implementation of this modification is not easy to do with a one-line anonymous function. Thus, we have created the function `makeMeasurementModel` within the `tracker` package. It will accept a set of coordinates for AOA, TDOA, and FDOA sensors, just as in the `hybrid` package, as well as the state_space struct defining how $\mathbf{x}$ is constructed, and will return two function handles, one for $\mathbf{z}(\mathbf{x})$ and another for $\mathbf{H}(\mathbf{x})$.

```
[z_fun, H_fun] = tracker.makeMeasurementModel(x_aoa, x_tdoa, x_fdoa, v_fdoa,
     tdoa_ref_idx, fdoa_ref_idx, state_space);
```

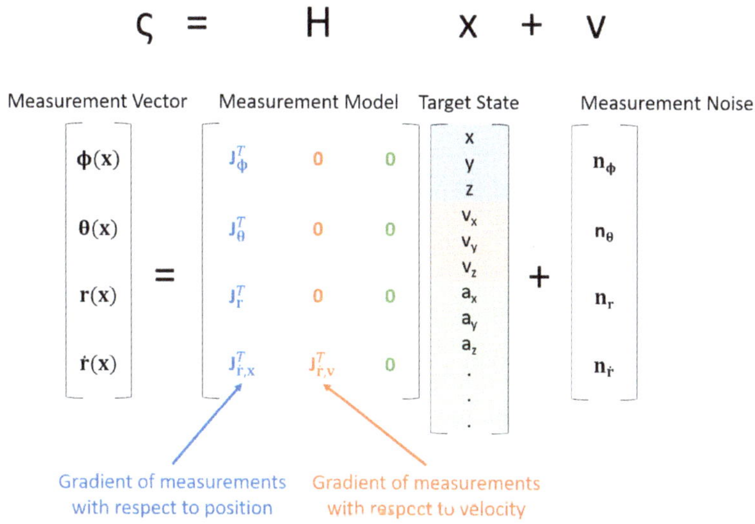

Figure 8.2 Illustration of the structure of the linearized measurement matrix **H**, and its relation to the Jacobian matrices of the individual measurement phenomenologies and the elements of target state **x**.

For other typical measurement models, not limited to those used for passive geolocation, see [3], [26]–[30].

Example 8.1 Tracking an Aircraft from Ground Sensors

Consider a 4-element TDOA system arranged as a triangle with receivers at (5, 0) km, (0, 0) km, (0, 5) km, and (−5, 0) km, the first two at an altitude of 30 meters and the latter two at 60 meters. Generate measurements from an aircraft that is initially at (−50, 100) km, at an altitude of 20 kft, and heading in the +x direction at 200 m/s. After 3 minutes, the aircraft begins a gradual level turn, with a radius of 50 km, maintaining speed through the turn, until is is headed in the $-y$ direction. It continues this way for 6 more minutes after exiting the turn.

When computing the aircraft's trajectory, make use of the rate of turn (ROT) in radians/second, which is computed for a constant velocity turn:

$$\text{ROT} = \frac{V}{R\pi/2} \tag{8.29}$$

where V is the speed, in meters/second, and R is the turn radius, in meters.

Every 10 seconds, the aircraft transmits a transponder pulse that is detected by the TDOA receivers, with a timing error of $\sigma_t = 10$ ns. Run an EKF tracker, using a CV model with acceleration noise of $\sigma_a = 1$ m/s^2. Initialize the state estimate of the track by using a GD solver on the first measurement, using an initial estimate of (0, 50e3, 5e3), and the position component of the state covariance with the CRLB on position error. For the velocity component of the state, assume an initial velocity of 0 m/s in all directions, but a standard deviation on the prior state of $\sigma_v = 340$ m/s. This corresponds to a Mach 1 aircraft (at sea level).

Plot the truth trajectory and estimated positions. Plot the CEP_{50} of both the prediction and estimation error, along with the true geolocation error for the prediction and estimation at each update.

Solution: The most complex part of this example is the definition of the trajectory, which we accomplish in four pieces. First, we define the transition times, including the start and end of the turn, and the end of the trajectory.

```
t_e_leg = 3*60; % time of eastbound leg
vel = 200; % m/s
turn_rad = 50e3;      % radius of the level turn
t_turn = pi/2*turn_rad / vel; % time to traverse a 1/4 arc
t_s_leg = 6*60; % time of southbound leg
```

Next, we define the trajectory for each of the three legs:

```
t_inc = 10; % 10 seconds between track updates
x_tgt_init = [-50e3;100e3;20e3*.3048];

x_e_leg = x_tgt_init + [vel;0;0]*(0:t_inc:t_e_leg);

angle_turn = pi/2 * (t_inc:t_inc:t_turn)/t_turn;
x_turn = x_e_leg(:,end) + turn_rad * [sin(angle_turn); cos(angle_turn)-1;
    zeros(size(angle_turn))];

x_s_leg = x_turn(:,end) + [0;-vel;0]*(t_inc:t_inc:t_s_leg);

x_tgt_full = cat(2,x_e_leg, x_turn, x_s_leg);
t_vec = 0:t_inc:(t_e_leg+t_turn+t_s_leg);
```

This trajectory is plotted in Figure 8.3. Once we have the target (and, similarly, the TDOA sensors) defined, we turn to the sensing performance. Here, we must be careful of the units, since many of the `tdoa` functions work in *range* difference rather than time difference.[4]

```
ref_idx = 1; % reference index for TDOA processing
sigma_toa = 10e-9; % timing error
C_toa = sigma_toa^2 * eye(n_tdoa); % TOA sensor-level covariance
C_roa = utils.constants.c^2 * C_toa; % ROA sensor-level covariance
```

4. When in doubt, consult the `help` interface, which will indicate what type of error units are accepted.

```
R = utils.resampleCovMtx(C_roa, ref_idx); % range-difference covariance
L = chol(R,'lower'); % noise generator
```

Now, we can generate the noisy measurements, one for each aircraft position.

```
z = tdoa.measurement(x_tdoa, x_tgt_full, ref_idx);
noise = L * randn(num_msmt, num_time);
zeta = z + noise;
```

Up until this point, we have not done anything tracker specific. Now, we must conduct our tracker analysis in three pieces. First, we'll initialize the model, then we'll initialize the track, and finally we'll iterate over the measurements and update the track with each.

```
sigma_a = 1; % process noise -- acceleration standard deviation
[f_fun, q_fun, state_space] = tracker.makeKinematicModel('cv',num_dims,
    sigma_a^2);
F = f_fun(t_inc);
Q = q_fun(t_inc);

[z_fun, h_fun] = tracker.makeMeasurementModel([],x_tdoa,[],[],ref_idx,[],
    state_space);
```

The reason that the kinematic model returns a function handle is to account for nonuniform measurement intervals (e.g. uncertainty as to when the next measurement will arrive). We now have the matrices `F` and `Q` that can be supplied to `kfPredict` and the function handles `z_fun` and `h_fun` that can be supplied to `ekfUpdate`.

Next, we initialize the track state. We begin by setting `x_pred` and `P_pred` to all zeros, then fill in what information we can. From our initial TDOA measurement, we can estimate target position, which we do with a GD solver, given an initial guess at (0, 50e3, 5e3), and then feed that initial position estimate into a CRLB calculation. Note that we use the term `pos_idx` and `vel_idx` to index into the appropriate entries of both the state and the error covariance.

```
pos_idx = state_space.pos_idx;
vel_idx = state_space.vel_idx;

x_pred = zeros(num_states,1);
P_pred = zeros(num_states, num_states);

% Initialize position with TDOA estimate from first measurement
x_init = [0;50e3;5e3];
epsilon = 100; % stopping condition
x_pred(pos_idx) = tdoa.gdSoln(x_tdoa, zeta(:,1), C_roa, x_init,[],[],epsilon
    ,[],[],[],ref_idx);
```

```
P_pred(pos_idx,pos_idx) = tdoa.computeCRLB(x_tdoa,x_pred(pos_idx),C_roa,
    ref_idx,false,true);

max_vel = 340;
P_pred(vel_idx, vel_idx) = max_vel^2*eye(num_dims);
```

Now that we have initialized the track, we can step through time and update it. What isn't shown here is that the predicted and estimated states (`x_pred` and `x_est`, respectively) are stored for later calculation of geolocation error and plotting the trajectory of estimates over time.[5]

```
for idx=1:num_time
    % Grab Current Measurement
    this_zeta = zeta(:,idx);

    % Update Position Estimate
    [x_est, P_est] = tracker.ekfUpdate(x_pred, P_pred, this_zeta, R, z_fun,
        h_fun);

    % Predict state to the next time step
    [x_pred, P_pred] = tracker.kfPredict(x_est, P_est, Q, F);
end
```

The estimates, along with the sensor positions and target velocity, are shown in Figure 8.3. The first few sensor updates have a large error but gradually converge to the track of the target, even following its turn without much difficulty. Had we set the random acceleration term too low (say, below .1 m/s^2), it would have failed to track the turn and could potentially have broken the track (if we had logic in our tracking loop that allowed for termination of a track after too many failed updates). If we set the random acceleration term too high (say, above 10 m/s^2), it would have slowed the convergence, and we may have missed the first straight leg portion of the track entirely before the track settled.

The error at each update, along with the RMSE computed from $\mathbf{P}_{k|k}$ (estimation) and $\mathbf{P}_{k|k-1}$ (prediction) at each update, are plotted in Figure 8.4. We look first at the dashed blue and red lines (estimation and prediction, respectively). They begin with errors on the order of 10 km and converge after approximately 250 seconds toward the predicted RMSE. In all cases, the difference between the predicted and estimated RMSE is due to the need to propagate an erroneous state estimate forward some period of time as well as the uncertainty due to process noise. Increasing the random acceleration term, or increasing the time duration between updates, would make the difference between predicted and estimated RMSE more stark.

For comparison, we also repeated the experiment for 1,000 Monte Carlo trials and reported the RMSE of geolocation error across time in Figure 8.5. We observe

5. In this snippet, we have excluded the code that stores `x_pred`, `x_est`, `P_pred`, and `P_est` across time steps for later plotting and error analysis.

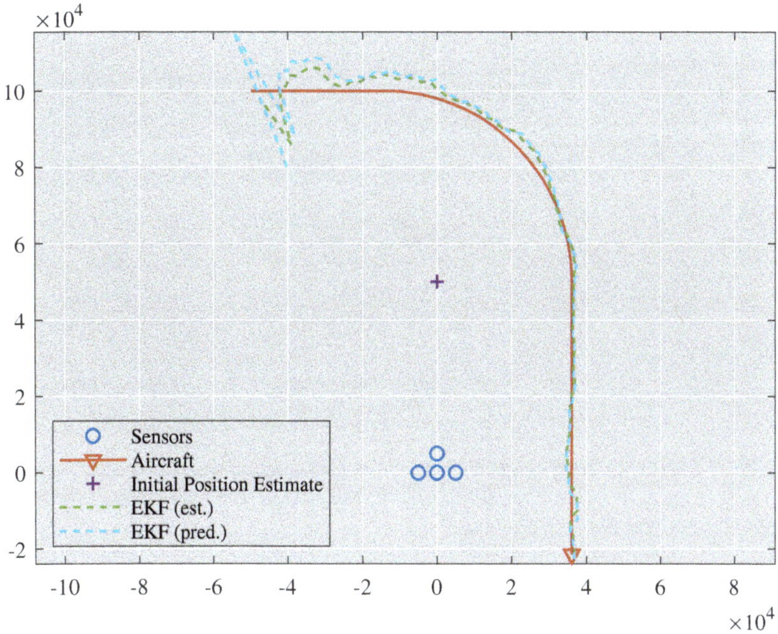

Figure 8.3 Trajectory and EKF tracker output for Example 8.1.

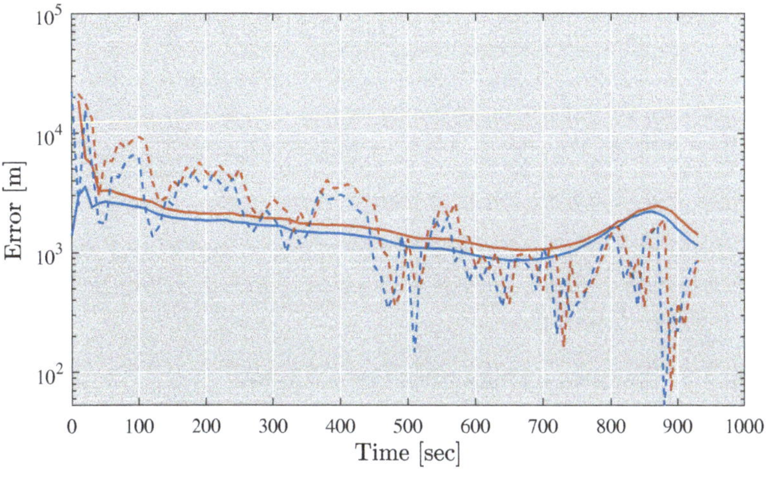

Figure 8.4 RMSE for both predicted and estimated state at each time step, compared with the CEP50 computed from the predicted and estimated state error covariance, for Example 8.1.

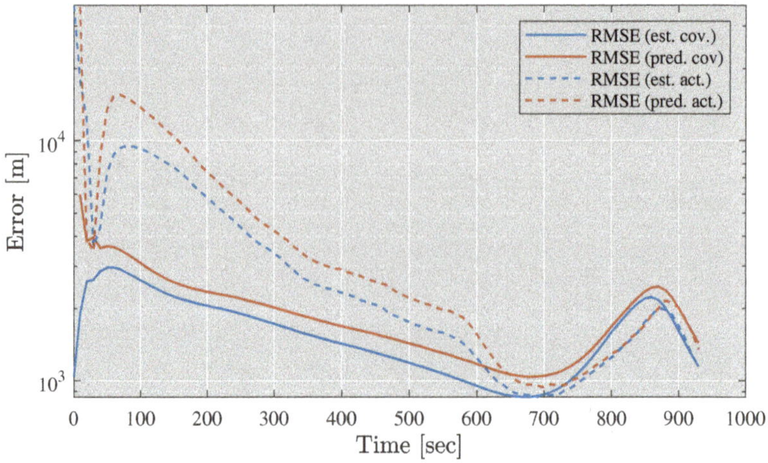

Figure 8.5 RMSE for both predicted and estimated state at each time step, compared with the CEP50 computed from the predicted and estimated state error covariance, for Example 8.1, averaged across 1,000 Monte Carlo trials.

that the error terms are all much smoother, as can be expected with any Monte Carlo trial, but also that there is a much longer convergence time as the achieved errors trend toward those reported. This is likely due to the presence of degenerate initial conditions. Depending on the initial measurement, we may start the track in a very poor state, from which it takes a larger number of updates to recover. One approach to minimize this risk is to use multiple measurements to initialize the track (as we will do in the next example). Also, we observe that there is an increase in error near the end of the test. This occurs when the target approaches the baseline for the three TDOA sensors located at $y = 0$, at roughly $t = 820$ s. This is a degenerate geometry for the arrangement of sensors chosen, and will result in reduced performance as the target approaches and passes through this point. □

Figure 8.6 contains a link to a video discussion of Example 8.1. In this video, we discuss in a little more detail the convergence of error and the use of the sample mean to improve sensor performance.

Example 8.2 Tracking a Ship from a Small Aircraft

Consider a small aircraft, that begins at the origin (0, 0), at an altitude of 10 kft. It is traveling in the +y direction at 100 m/s. A ship begins at the point (50, 50) km, headed in the $-x$ direction at a speed of 10 m/s. The ship is broadcasting a transponder signal every 30 seconds, which is detected by the small aircraft, with an angle accuracy of $\sigma_\theta = 1°$ in azimuth and elevation.

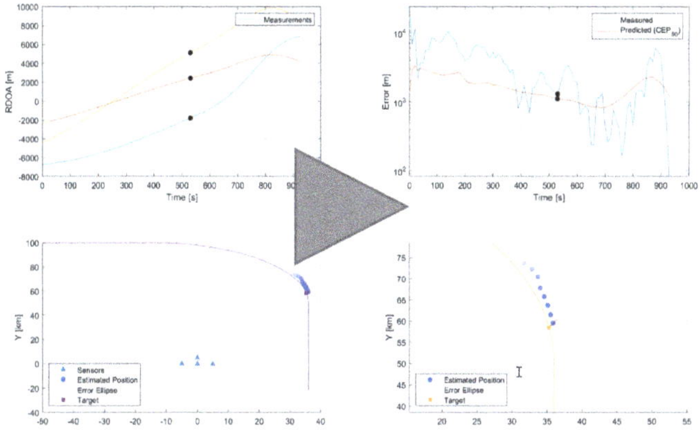

Figure 8.6 Video discussing the convergence of sample measurements in Example 8.1.

To account for the impact of wave action, generate a random acceleration term for the ship that is uniformly distributed between $-.05$ m/s^2, and $.05$ m/s^2, in both the x and y dimensions.

Generate a set of random measurements over a period of 15 minutes. Use the first three measurements jointly to initialize the track (ignoring target motion). Use a maximum velocity of 30 m/s when initializing the track error covariance. Use knowledge of the target's position on the surface of the ocean to improve geolocation performance. For a kinematic model, use the constant velocity model, with a random acceleration power of $\sigma_a = .05$ m/s^2.

Plot the layout, tracker results, and geolocation error.

Solution: To begin, we define the aircraft and target positions. For the aircraft, the approach is very straightforward. The ship is slightly more complex. We begin by defining an acceleration power and generating random uniform acceleration on the prescribed interval. We then integrate (via a `cumsum` operation) the random acceleration to generate velocity at each time step, and repeat the process for position. These two trajectories are plotted in Figure 8.7(a), along with a few example AOA measurements (with a 1σ confidence interval).

```
x_aoa_full = [0;0;10e3] + [0;100;0] * t_vec;
v_aoa_full = [0;100;0]*ones(size(t_vec)); % constant vel

ship_accel_power = .05;
a_tgt_full = cat(1,-ship_accel_power + 2*ship_accel_power*rand(2,num_time),
    zeros(1,num_time));
v_tgt_full = [-20;0;0] + cumsum(a_tgt_full*t_inc,2);
x_tgt_full = [50e3;50e3;0] + cumsum(v_tgt_full*t_inc,2);
```

Next, we define the measurement statistics and generate a set of noise measurements, in the same manner as has been done for all of our examples.

```
sigma_theta = 1; % deg
sigma_psi = sigma_theta*pi/180; % rad
num_msmt=2; % az/el
R = sigma_psi^2*eye(num_msmt); % measurement error covariance

z = zeros(num_msmt,num_time);
for idx=1:num_time
    z(:,idx) = triang.measurement(x_aoa_full(:,idx),x_tgt_full(:,idx),true);
end
noise = sigma_psi * randn(num_msmt, num_time);
zeta = z + noise;
```

Next, we set up the tracker's kinematic model and then initialize the track state and error, just as in the previous model. In this case, however, we supply the first three measurements to a constrained gradient descent solver to take advantage of knowledge that the target is on the Earth's surface. We use this information again when initializing the error covariance estimate.

```
sigma_a = .05;
num_dims = 3; % number of dimensions to use in state

% Kinematic Model
[f_fun, q_fun, state_space] = tracker.makeKinematicModel('cv',num_dims,
    sigma_a^2);
num_states = state_space.num_states;
pos_idx = state_space.pos_idx;
vel_idx = state_space.vel_idx;
F = f_fun(t_inc); % generate state transition matrix
Q = q_fun(t_inc); % generate process noise covariance matrix

x_pred = zeros(num_states,1);
P_pred = zeros(num_states, num_states);

% Initialize position with AOA estimate from first three measurement
x_aoa_init = x_aoa_full(:,1:3); % sensor positions
z_init = reshape(z(:,1:3)',[],1); % collect the 3 az, then the 3 el msmts
C_init = kron(R,eye(3)); % expand the covariance error

x_init_guess = [10e3;10e3;0];
[bnd, bnd_grad] = utils.constraints.fixedAlt(0,'flat');
x_init = triang.gdSolnFixed(x_aoa_init,z_init,C_init,x_init_guess,bnd
    ,100,[],[],1e3,[],[],[]);
P_init = triang.computeCRLBfixed(x_aoa_init, x_init, C_init, bnd_grad, true)
    ;
```

```
% Update state and error estimates for position
x_pred(pos_idx) = x_init(1:num_dims);
P_pred(pos_idx, pos_idx) = P_init(1:num_dims,1:num_dims);

% Bound initial velocity uncertainty by assumed max velocity in
% x and y (z velocity is assumed known perfectly to be zero)
max_vel = 30;
P_pred(vel_idx(1:2), vel_idx(1:2)) = max_vel^2*eye(num_dims);
```

The tracking loop is the same as in Example 8.1, with two notable changes. First, we recompute the measurement model at each time step to account for sensor motion. Second, the estimates are manually adjusted to reflect knowledge of the target's altitude. This is not as elegant as the constrained solvers presented in Chapter 5, but does suffice as an initial attempt at constrained target tracking.

```
for idx=1:num_time
    % Grab Current Measurement and update measurement function
    this_zeta = zeta(:,idx);
    this_x_aoa = x_aoa_full(1:num_dims,idx);
    [z_fun, h_fun] = tracker.makeMeasurementModel(this_x_aoa,[],[],[],[],[],
        state_space);

    % Update Position Estimate
    [x_est, P_est] = tracker.ekfUpdate(x_pred, P_pred, this_zeta, R, z_fun,
        h_fun);

    % Constrain vertical position and velocity
    x_est(pos_idx(end)) = 0;
    x_est(vel_idx(end)) = 0;
    P_est(pos_idx(end),:) = 0;
    P_est(:,pos_idx(end)) = 0;
    P_est(vel_idx(end),:) = 0;
    P_est(:,vel_idx(end)) = 0;

    % Predict state to the next time step
    [x_pred, P_pred] = tracker.kfPredict(x_est, P_est, Q, F);

        % Constrain vertical position and velocity
    x_pred(pos_idx(end)) = 0;
    x_pred(vel_idx(end)) = 0;
    P_pred(pos_idx(end),:) = 0;
    P_pred(:,pos_idx(end)) = 0;
    P_pred(vel_idx(end),:) = 0;
    P_pred(:,vel_idx(end)) = 0;
end
```

We plot the geometry and both the predicted and estimated positions over time in Figure 8.7. Figure 8.7(b) shows a close-up image of the target trajectory, from which we see that the initial position estimates are gradually improved, as the track converges to a better track of the target. This may appear worse than the result in Example 8.1, but we note that the slower velocity of the target in this exaggerates the appearance of *jumpiness* in our track. To see for sure how accurate the track is, we'll analyze geolocation error as a function of time.

Figure 8.8 plots the error over time for both prediction and estimation at each step, as well as the predicted error from the error covariance matrix estimates. The results agree with our impression from the geographic plot. Initial errors are on the order of 3-4 km, but they converge to slightly better than 1 km by the end of the 15-minute observation period. Just as before, we see that as the aircraft flies away from the ship, the predicted error performance starts to increase again, shown by the orange and blue solid lines, which begin to trend upward after 700 seconds. □

Figure 8.9 contains a link to a video discussion of Example 8.2. In this video, we discuss in a little more detail the convergence of error and the use of the sample mean to improve sensor performance.

8.4 ADVANCED TRACKING CONCEPTS

In this chapter, we have introduced the concept of a tracker and discussed how it can be used to perform geolocation of moving targets with passive sensor measurements. By necessity, our discussion of tracking is cursory. We have covered the core mathematical concepts of prediction and updating estimates as measurements are made over time, but have not discussed any of the many critical auxiliary processes and algorithms that are required to assemble a working tracker. A comprehensive treatment of many of these practicalities can be found in [11].

For example, in any realistic scenario there are false alarms, missed detections, and multiple targets. In such an environment, the first step to tracking is to associate incoming measurements with a track, or to declare them as new tracks (*data association*). This leads to the need for logic to determine when a new track should be formed (*track initiation*) and when a stale track should be deleted (*track deletion*).

There are many approaches to track association, including *nearest neighbor*, in which each new measurement is assigned to the track whose predicted state most closely aligns with the measurement [11]. Performance can be increased in complex scenarios with *probabilistic data association* (PDA) [31]–[33], a greedy algorithm similar to nearest neighbor that also considers prediction error covariance, and the related *joint probabilistic data association* (JPDA) [34]–[36], which optimizes jointly over all measurement-track pairings.

In especially complex scenarios, it may not be obvious which association decision is correct. *Multihypothesis tracking* (MHT) is a class of trackers that allow for multiple

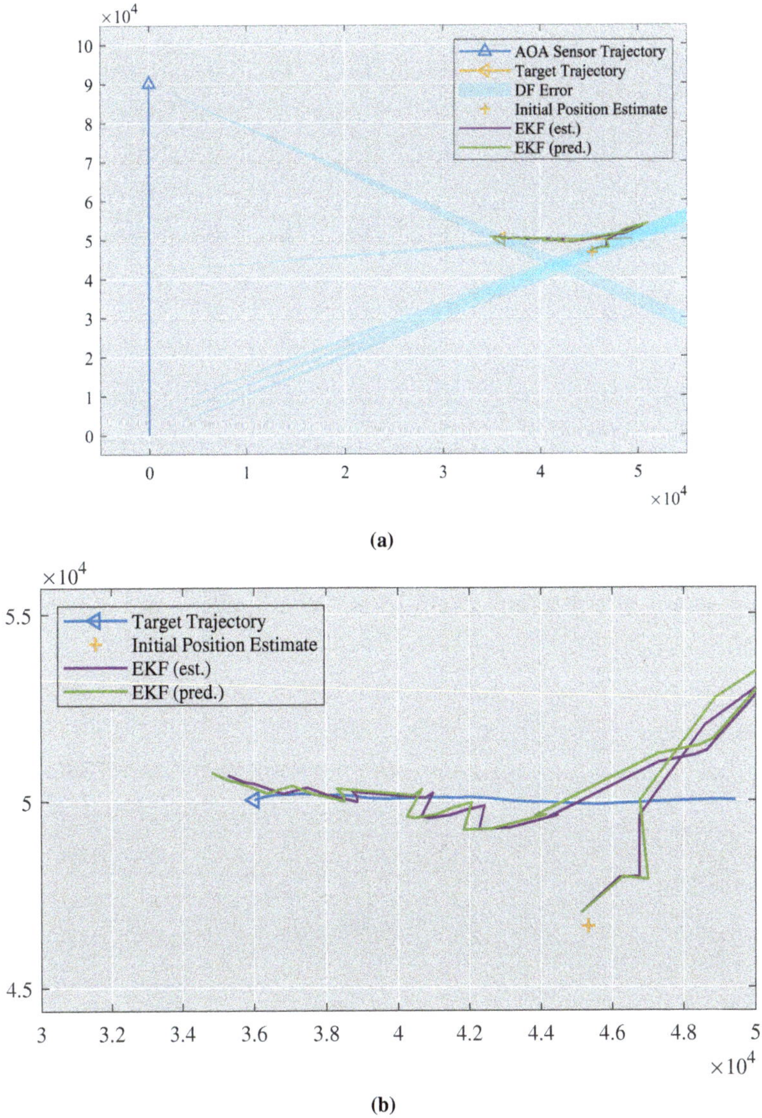

Figure 8.7 Layout of sensors and target, and tracker output, for the ship tracking scenario in Example 8.2. (a) Full scenario, and (b) target area.

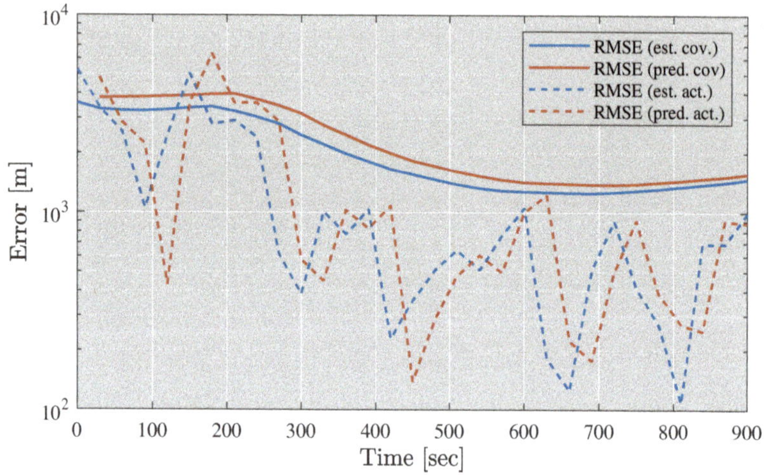

Figure 8.8 Plot of the error over time for the tracker in Example 8.2.

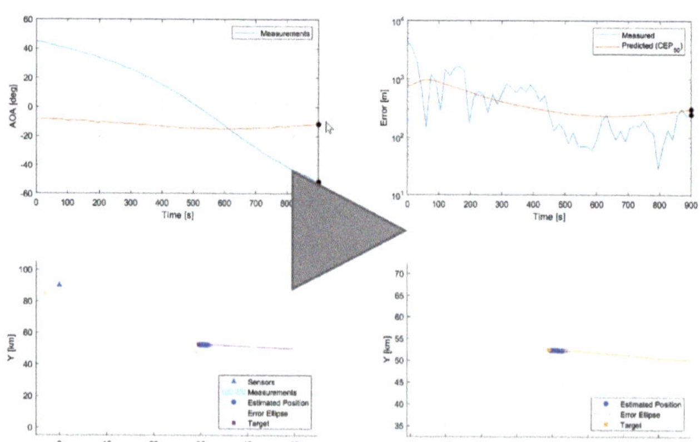

Figure 8.9 Video discussing the convergence of sample measurements in Example 8.2.

data association assignments to be hypothesized [37]–[39]. They each run in parallel, and the decision over which assignment is correct is deferred until later. The premise of this approach is that incorrect associations will lead to malformed tracks that will poorly predict future measurements and fail to converge, while the correct hypothesis will continue to improve. MHT is very powerful but can be incredibly computationally intensive. It has been applied to radar [40]–[42] and other RF tracking problems [43]–[45]. Proper pruning of infeasible hypotheses is key to success.

We noted in Section 8.3.2 that there are many types of kinematic models and introduced some of the more popular ones, along with situations in which each is likely to perform well. A principal challenge, of course, is understanding which situation is relevant. One approach, similar to the MHT, called *interacting multiple models* (IMM), is to run multiple kinematic models in parallel, periodically deciding which model was relevant for the previous section of track updates [20]. In this manner, a tracker may choose the CV model when a target is believed to be in level flight, the CA model when a turn is detected, and the CJ model when the target engages in more dynamic maneuvers. Similarly, the assumptions of process noise could be changed between models. The principal limitation to IMM is computational complexity [20]–[24].

Finally, we discussed two straightforward tracker formulations, the KF and EKF. There is a third common extension of the KF, called the *unscented Kalman filter* (UKF), which provides improved performance over the EKF for highly nonlinear transition or measurement models (those where the first-order Taylor series approximation in the EKF is not sufficiently accurate). The UKF uses a series of *sigma points* around the predicted/estimated target state, propagates each of these through the nonlinear function, and then estimates a new mean and covariance from the outputs [46]–[48].

Another common class of tracker is the *particle filter*, sometimes referred to as *sequential Monte Carlo* filters [49], [50]. Particle filters use random perturbations to estimate the dynamics of a nonlinear system. A particular advantage, with respect to the KF, EKF, and UKF described earlier, is that particle filters do not rely on a Gaussian assumption to process and measurement noise terms.

The *modified gain EKF* (MGEKF) uses the innovation between successive prediction and estimation stages to more accurately update track error covariance matrix $\mathbf{P}_{k|k}$ [51], and was the basis for a modified bearing-only tracker from Guerci et al. that has an improved initialization step, as well [52]. Interested readers are referred to these sources as a first step into the voluminous research literature available on specialized trackers.

While the tracking code provided in this text is well integrated into the rest of our geolocation code, readers may be interested in the open source Python library *Stone Stoup*[53], which aims to provide a library of building blocks for assembling complex tracking systems.

8.5 PROBLEM SET

8.1 What kinematic model, from among those listed in Table 8.1, would you select to track a maneuvering target that is executing a series of weaving patterns, maintaining a steady speed and rate of turn in each. Describe your choice.

8.2 Why would you choose a constant jerk kinematic model over constant acceleration? What are the drawbacks of doing so?

8.3 Repeat Example 8.1, but with FDOA sensors instead of TDOA. Assume a carrier frequency of 600 MHz and a frequency estimate error of 10 Hz.

8.4 Repeat Example 8.1, but move the left receiver 5 km to the left, and the right receiver 5 km down. Comment on the differences in tracking accuracy.

References

[1] R. E. Kalman, "A new approach to linear filtering and prediction problems," *Transactions of the ASME–Journal of Basic Engineering*, 1960.

[2] L. L. Scharf, *Statistical Signal Processing*. Reading, MA: Addison-Wesley, 1991.

[3] K. Spingarn, "Passive position location estimation using the extended Kalman filter," *IEEE Transactions on Aerospace and Electronic Systems*, vol. AES-23, no. 4, pp. 558–567, 1987. DOI: 10.1109/TAES.1987.310889.

[4] R. Faragher, "Understanding the basis of the Kalman filter via a simple and intuitive derivation [lecture notes]," *IEEE Signal Processing Magazine*, vol. 29, no. 5, pp. 128–132, 2012. DOI: 10.1109/MSP.2012.2203621.

[5] P. Hargrave, "A tutorial introduction to Kalman filtering," *IEE Colloquium on Kalman Filters: Introduction, Applications and Future Developments*, pp. 1/1–1/6, 1989.

[6] D. P. Mandic, S. Kanna, and A. G. Constantinides, "On the intrinsic relationship between the least mean square and Kalman filters [lecture notes]," *IEEE Signal Processing Magazine*, vol. 32, no. 6, pp. 117–122, 2015. DOI: 10.1109/MSP.2015.2461733.

[7] J. Humpherys and J. West, "Kalman filtering with Newton's method [lecture notes]," *IEEE Control Systems Magazine*, vol. 30, no. 6, pp. 101–106, 2010. DOI: 10.1109/MCS.2010.938485.

[8] A. Kheirati Roonizi, "An efficient algorithm for maneuvering target tracking [tips amp; tricks]," *IEEE Signal Processing Magazine*, vol. 38, no. 1, pp. 122–130, 2021. DOI: 10.1109/MSP.2020.3029386.

[9] M. S. Grewal and A. P. Andrews, *Kalman Filtering: Theory and Practice with MATLAB*. John Wiley & Sons, 2014.

[10] F. Daum, "Nonlinear filters: Beyond the Kalman filter," *IEEE Aerospace and Electronic Systems Magazine*, vol. 20, no. 8, pp. 57–69, 2005. DOI: 10.1109/MAES.2005.1499276.

[11] C.-B. Chang and K.-P. Dunn, *Applied State Estimation and Association*. MIT press, 2016.

[12] N. A. O'Donoughue, *Emitter Detection and Geolocation for Electronic Warfare*. Norwood, MA: Artech House, 2019.

[13] R. Kalman, "On the general theory of control systems," *IRE Transactions on Automatic Control*, vol. 4, no. 3, pp. 110–110, 1959.

[14] C. Chang, "Ballistic trajectory estimation with angle-only measurements," *IEEE Transactions on Automatic Control*, vol. 25, no. 3, pp. 474–480, 1980. DOI: 10.1109/TAC.1980.1102368.

[15] C. B. Chang and K.-P. Dunn, "Radar tracking using state estimation and association," in *2019 IEEE Radar Conference (RadarConf)*, 2019, pp. 1–220. DOI: 10.1109/RADAR.2019.8835799.

[16] X. Rong Li and V. Jilkov, "Survey of maneuvering target tracking. Part I. Dynamic models," *IEEE Transactions on Aerospace and Electronic Systems*, vol. 39, no. 4, pp. 1333–1364, 2003. DOI: 10.1109/TAES.2003.1261132.

[17] K. Mehrotra and P. R. Mahapatra, "A jerk model for tracking highly maneuvering targets," *IEEE Transactions on Aerospace and Electronic Systems*, vol. 33, no. 4, pp. 1094–1105, 1997.

[18] X. R. Li and V. P. Jilkov, "Survey of Maneuvering Target Tracking. Part II: Motion Models of Ballistic and Space Targets," *IEEE Transactions on Aerospace and Electronic Systems*, vol. 46, no. 1, pp. 96–119, 2010. DOI: 10.1109/TAES.2010.5417150.

[19] C.-B. Chang, M. Athans, and R. Whiting, "On the state and parameter estimation for maneuvering reentry vehicles," *IEEE Transactions on Automatic Control*, vol. 22, no. 1, pp. 99–105, 1977. DOI: 10.1109/TAC.1977.1101412.

[20] X. Rong Li and V. Jilkov, "Survey of maneuvering target tracking. Part V. Multiple-model methods," *IEEE Transactions on Aerospace and Electronic Systems*, vol. 41, no. 4, pp. 1255–1321, 2005. DOI: 10.1109/TAES.2005.1561886.

[21] E. Mazor, A. Averbuch, Y. Bar-Shalom, and J. Dayan, "Interacting multiple model methods in target tracking: A survey," *IEEE Transactions on Aerospace and Electronic Systems*, vol. 34, no. 1, pp. 103–123, 1998. DOI: 10.1109/7.640267.

[22] X. Li and Y. Bar-Shalom, "Performance prediction of the interacting multiple model algorithm," *IEEE Transactions on Aerospace and Electronic Systems*, vol. 29, no. 3, pp. 755–771, 1993. DOI: 10.1109/7.220926.

[23] X. Li and Y. Bar-Shalom, "Design of an interacting multiple model algorithm for air traffic control tracking," *IEEE Transactions on Control Systems Technology*, vol. 1, no. 3, pp. 186–194, 1993. DOI: 10.1109/87.251886.

[24] A. Munir and D. Atherton, "Adaptive interacting multiple model algorithm for tracking a manoeuvring target," *IEE Proceedings - Radar, Sonar and Navigation*, vol. 142, 11–17(6), 1 Feb. 1995, ISSN: 1350-2395. [Online]. Available: https://digital-library.theiet.org/content/journals/10.1049/ip-rsn_19951528.

[25] K. Saho and M. Masugi, "Automatic parameter setting method for an accurate Kalman filter tracker using an analytical steady-state performance index," *IEEE Access*, vol. 3, pp. 1919–1930, 2015. DOI: 10.1109/ACCESS.2015.2486766.

[26] X. R. Li and V. P. Jilkov, "Survey of maneuvering target tracking: III. Measurement models," *Signal and Data Processing of Small Targets 2001*, vol. 4473, pp. 423–446, 2001. DOI: 10.1117/12.492752. [Online]. Available: https://doi.org/10.1117/12.492752.

[27] I. Leibowicz, P. Nicolas, and L. Ratton, "Radar/ESM tracking of constant velocity target: comparison of batch (MLE) and EKF performance," *Proc. of the Third International Conference on Information Fusion*, vol. 1, pp. 3–8, 2000. DOI: 10.1109/IFIC.2000.862680.

[28] A. Farina, "Target tracking with bearings – only measurements," *Signal Processing*, vol. 78, no. 1, pp. 61–78, 1999, ISSN: 0165-1684. DOI: https://doi.org/10.1016/S0165-1684(99)00047-X. [Online]. Available: https://www.sciencedirect.com/science/article/pii/S016516849900047X.

[29] L. Ratton and A. Tournade, "Track maintenance in covert mission: Passive nonlinear filtering through Cartesian vs modified polar EKF," *2012 15th International Conference on Information Fusion*, pp. 2079–2086, 2012.

[30] M. Malanowski, *Signal Processing for Passive Bistatic Radar*. Norwood, MA: Artech House, 2019.

[31] Y. Bar-Shalom and E. Tse, "Tracking in a cluttered environment with probabilistic data association," *Automatica*, vol. 11, no. 5, pp. 451–460, 1975.

[32] T. Kirubarajan and Y. Bar-Shalom, "Probabilistic data association techniques for target tracking in clutter," *Proceedings of the IEEE*, vol. 92, no. 3, pp. 536–557, 2004. DOI: 10.1109/JPROC.2003.823149.

[33] Y. Bar-Shalom, T. Kirubarajan, and X. Lin, "Probabilistic data association techniques for target tracking with applications to sonar, radar and eo sensors," *IEEE Aerospace and Electronic Systems Magazine*, vol. 20, no. 8, pp. 37–56, 2005. DOI: 10.1109/MAES.2005.1499275.

[34] T. E. Fortmann, Y. Bar-Shalom, and M. Scheffe, "Multi-target tracking using joint probabilistic data association," *Proc. 19th IEEE Conference on Decision and Control including the Symposium on Adaptive Processes*, pp. 807–812, 1980.

[35] T. Fortmann, Y. Bar-Shalom, and M. Scheffe, "Sonar tracking of multiple targets using joint probabilistic data association," *IEEE Journal of Oceanic Engineering*, vol. 8, no. 3, pp. 173–184, 1983. DOI: 10.1109/JOE.1983.1145560.

[36] K.-C. Chang and Y. Bar-Shalom, "Joint probabilistic data association for multitarget tracking with possibly unresolved measurements and maneuvers," *IEEE Transactions on Automatic Control*, vol. 29, no. 7, pp. 585–594, 1984. DOI: 10.1109/TAC.1984.1103597.

[37] D. Reid, "An algorithm for tracking multiple targets," *IEEE Transactions on Automatic Control*, vol. 24, no. 6, pp. 843–854, 1979. DOI: 10.1109/TAC.1979.1102177.

[38] R. L. Streit and T. E. Luginbuhl, "Probabilistic multi-hypothesis tracking," NAVAL UNDERWATER SYSTEMS CENTER NEWPORT RI, Tech. Rep. ADA298501, 1995. [Online]. Available: https://apps.dtic.mil/sti/citations/ADA298501.

[39] Y. Bar-Shalom, *Multitarget-Multisensor Tracking: Advanced Applications*. Norwood, MA: Artech House, 1990.

[40] S. S. Blackman and R. Popoli, *Design and analysis of modern tracking systems*. Norwood, MA: Artech House, 1999.

[41] R. F. Popoli, S. S. Blackman, and M. T. Busch, "Application of multiple-hypothesis tracking to agile beam radar tracking," *Signal and Data Processing of Small Targets*, pp. 418–428, 1996. DOI: 10.1117/12.241203. [Online]. Available: https://doi.org/10.1117/12.241203.

[42] M. Hernandez, A. Benavoli, A. Graziano, A. Farina, and M. Morelande, "Performance measures and mht for tracking move-stop-move targets with mti sensors," *IEEE Transactions on Aerospace and Electronic Systems*, vol. 47, no. 2, pp. 996–1025, 2011. DOI: 10.1109/TAES.2011.5751239.

[43] J. S. Peri, "Approaches to multisensor data fusion," *Johns Hopkins APL Technical Digest*, vol. 22, no. 4, pp. 624–633, 2001.

[44] R. J. Dempster, S. S. Blackman, S. H. Roszkowski, and D. M. Sasaki, "IMM/MHT solution to radar multisensor benchmark tracking problems," *Signal and Data Processing of Small Targets 1998*, pp. 331–342, 1998. DOI: 10.1117/12.324629. [Online]. Available: https://doi.org/10.1117/12.324629.

[45] S. Blackman, R. Dempster, M. Busch, and R. Popoli, "IMM/MHT solution to radar benchmark tracking problem," *IEEE Transactions on Aerospace and Electronic Systems*, vol. 35, no. 2, pp. 730–738, 1999. DOI: 10.1109/7.766953.

[46] S. J. Julier and J. K. Uhlmann, "New extension of the Kalman filter to nonlinear systems," *Signal processing, sensor fusion, and target recognition VI*, pp. 182–193, 1997.

[47] H. M. T. Menegaz, J. Y. Ishihara, G. A. Borges, and A. N. Vargas, "A systematization of the unscented kalman filter theory," *IEEE Transactions on Automatic Control*, vol. 60, no. 10, pp. 2583–2598, 2015. DOI: 10.1109/TAC.2015.2404511.

[48] W. Li, H. Leung, and Y. Zhou, "Space-time registration of radar and ESM using unscented Kalman filter," *IEEE Transactions on Aerospace and Electronic Systems*, vol. 40, no. 3, pp. 824–836, 2004. DOI: 10.1109/TAES.2004.1337457.

[49] N. J. Gordon, D. J. Salmond, and A. F. Smith, "Novel approach to nonlinear/non-Gaussian Bayesian state estimation," *IEE Proceedings F-radar and signal processing*, vol. 140, no. 2, pp. 107–113, 1993.

[50] F. Gustafsson, "Particle filter theory and practice with positioning applications," *IEEE Aerospace and Electronic Systems Magazine*, vol. 25, no. 7, pp. 53–82, 2010. DOI: 10.1109/MAES.2010.5546308.

[51] T. Song and J. Speyer, "A stochastic analysis of a modified gain extended kalman filter with applications to estimation with bearings only measurements," *IEEE Transactions on Automatic Control*, vol. 30, no. 10, pp. 940–949, 1985.

[52] J. Guerci, R. Goetz, and J. DiModica, "A method for improving extended Kalman filter performance for angle-only passive ranging," *IEEE Transactions on Aerospace and Electronic Systems*, vol. 30, no. 4, pp. 1090–1093, 1994. DOI: 10.1109/7.328754

[53] S. S. Contributors, *Stone soup*. [Online]. Available: https://stonesoup.readthedocs.io/en/latest/.

Appendix A

Derivation of the Fisher Information Matrix for Phase, Delay, and Doppler Estimation

This appendix is a reproduction of material from Chapter 13 of [1] and is reproduced here to provide a derivation of the Fisher information matrix for joint estimation of a signal's phase shift, time delay, and Doppler shift, from the measurement of the signal at two locations. It is used in (2.22).

The direct estimation of frequency difference (Doppler shift for narrowband signals) has been widely studied under various conditions and various assumptions on the known/unknown nuisance parameters, such as amplitude, delay, and phase differences between the two signals, and whether one of them is noise-free (i.e., a known reference signal, or a second noisy received signal treated as a reference). Interested readers are directed to [2]–[8]

Estimation of time and frequency is most easily computed via the cross-ambiguity function [9], [10]. Given complex data from sensors n and m, we define the cross-ambiguity function (for some frequency shift ω and delay τ)

$$\chi_{n,m}\left(\omega, \tau\right) = \int_0^T s_n(t) s_m^*\left(t + \tau\right) e^{-\jmath\omega t} dt \tag{A.1}$$

The estimate of time and frequency shift is given by searching for the peak of $\chi_{n,m}\left(\omega, \tau\right)$.

$$\left[\widehat{\omega}_{n,m}, \widehat{\tau}_{n,m}\right] = \arg\max_{\omega,\tau} \left|\chi_{n,m}\left(\omega, \tau\right)\right| \tag{A.2}$$

The classical result for Cramér-Rao bound on TDOA and FDOA between two receivers is given by Stein in 1981 [11] but was derived for a sonar signal model, in which the variance of the received signal is dependent on the unknown relative velocity. In a proper RF signal model, the delay and Doppler are deterministic (but unknown) parameters, so they have no impact on signal variance, and a much tighter bound can be formed [9], [12]. In the signal model below, delay (κ) is the digital delay in samples

($\tau = \kappa t_s$ for sample rate t_s), and doppler (ν) is in radians/sample ($\omega = \nu / t_s$).

$$\mathbf{s}_1 = \mathbf{s} + \mathbf{n}_1 \tag{A.3}$$
$$\mathbf{s}_2 = a e^{\jmath\phi}\mathbf{D}_\nu\left[\mathbf{W}^H\mathbf{D}_\kappa\mathbf{W}\right]\mathbf{s} + \mathbf{n}_2 \tag{A.4}$$

where $\mathbf{F}$ is a unitary DFT matrix, $\mathbf{D}_\kappa$ is the *delay* matrix and $\mathbf{D}_\nu$ is the *Doppler* matrix, defined as

$$\mathbf{W} = \frac{1}{\sqrt{M}}\exp\left(-\jmath\frac{2\pi}{M}\mathbf{m}\mathbf{m}^T\right) \tag{A.5}$$
$$\mathbf{D}_\kappa = \text{diag}\left\{\exp\left(-\jmath\frac{2\pi}{M}\kappa\mathbf{m}\right)\right\} \tag{A.6}$$
$$\mathbf{D}_\nu = \text{diag}\left\{\exp\left(-\jmath\nu\mathbf{m}\right)\right\} \tag{A.7}$$
$$\mathbf{m} = \left[-\frac{M}{2}, -\frac{M}{2}+1, \dots, \frac{M}{2}-1\right]^T \tag{A.8}$$

The full unknown parameter vector is

$$\boldsymbol{\vartheta} = \left[\Re\left\{\mathbf{s}^T\right\}, \Im\left\{\mathbf{s}^T\right\}, a, \phi, \kappa, \nu\right]^T \tag{A.9}$$

It can be shown that the FIM is block diagonal, with an upper subblock for the unknown signal ($\mathbf{s}$) and the amplitude difference (a), and a lower subblock for the phase (ϕ), delay (κ), and doppler (ν). The cross-terms between these blocks are all zero, so when we invert, the lower diagonal subblock (for $\overline{\boldsymbol{\vartheta}} = [\phi, \kappa, \nu]^T$) is not impacted by the upper subblock. Thus, we can ignore the uncertainty in estimating both $\mathbf{s}$ and a. The second subblock is given [9]

$$\mathbf{F}_{\phi,\kappa,\nu} = \frac{2}{a^2\sigma_1^2 + \sigma_2^2}\begin{bmatrix} E_s & -\mathbf{s}^H\overline{\mathbf{s}} & \widetilde{\mathbf{s}}^H\mathbf{M}\widetilde{\mathbf{s}} \\ -\mathbf{s}^H\overline{\mathbf{s}} & \overline{\mathbf{s}}^H\overline{\mathbf{s}} & -\Re\left\{\overline{\mathbf{s}}^H\mathbf{Q}_{\kappa,\nu}^H\mathbf{M}\widetilde{\mathbf{s}}\right\} \\ \widetilde{\mathbf{s}}^H\mathbf{M}\widetilde{\mathbf{s}} & -\Re\left\{\overline{\mathbf{s}}^H\mathbf{Q}_{\kappa,\nu}^H\mathbf{M}\widetilde{\mathbf{s}}\right\} & \widetilde{\mathbf{s}}^H\mathbf{M}^2\widetilde{\mathbf{s}}\end{bmatrix} \tag{A.10}$$

where σ_1^2 and σ_2^2 are the noise variances on $s_1(t)$ and $s_2(t)$, respectively, and the modified signal vectors and auxiliary matrices are defined as

$$\overline{\mathbf{s}} = \frac{2\pi}{M}\mathbf{W}^H\mathbf{M}\mathbf{W}\mathbf{s} \tag{A.11}$$
$$\widetilde{\mathbf{s}} = \mathbf{Q}_{\kappa,\nu}\mathbf{s} = \mathbf{D}_\nu\left[\mathbf{W}^H\mathbf{D}_\kappa\mathbf{W}\right]\mathbf{s} \tag{A.12}$$
$$\mathbf{M} = \text{diag}\left\{\mathbf{m}\right\} \tag{A.13}$$
$$\mathbf{Q}_{\kappa,\nu} = \mathbf{D}_\nu\left[\mathbf{W}^H\mathbf{D}_\kappa\mathbf{W}\right] \tag{A.14}$$

We note that the diagonal terms are the signal energy, a term that resembles RMS bandwidth (multiplied by signal energy), and a term that resembles RMS bandwidth (multiplied by signal energy). If we were to ignore the off-diagonal terms, then inverting the remaining matrix would result in the bounds on DF, TDOA, and FDOA accuracy from Chapters 11–13 of [1] and reproduced concisely in Section 2.4.5 of this text.

References

[1] N. A. O'Donoughue, *Emitter Detection and Geolocation for Electronic Warfare*. Norwood, MA: Artech House, 2019.

[2] C. H. Knapp and G. C. Carter, "Estimation of time delay in the presence of source or receiver motion," *The Journal of the Acoustical Society of America*, vol. 61, no. 6, pp. 1545–1549, 1977.

[3] E. Weinstein and P. M. Schultheiss, "Localization of a moving source using passive array data," Naval Ocean System Center Technical Report, Tech. Rep., 1978.

[4] P. M. Schultheiss and E. Weinstein, "Estimation of differential Doppler shifts," *The Journal of the Acoustical Society of America*, vol. 66, no. 5, pp. 1412–1419, 1979.

[5] M. Wax, "The joint estimation of differential delay, Doppler, and phase," *IEEE Transactions on Information Theory*, vol. 28, no. 5, pp. 817–820, Sep. 1982, ISSN: 0018-9448. DOI: 10.1109/TIT.1982.1056563.

[6] B. Friedlander, "On the Cramer-Rao bound for time delay and Doppler estimation," *IEEE Transactions on Information Theory*, vol. 30, no. 3, pp. 575–580, 1984.

[7] D. Mušicki and W. Koch, "Geolocation using TDOA and FDOA measurements," *Proc. 11th International Conference on Information Fusion*, pp. 1–8, Jun. 2008.

[8] B. Friedlander, "An efficient parametric technique for Doppler-delay estimation," *IEEE Transactions on Signal Processing*, vol. 60, no. 8, pp. 3953–3963, Aug. 2012, ISSN: 1053-587X. DOI: 10.1109/TSP.2012.2198818.

[9] A. Yeredor and E. Angel, "Joint TDOA and FDOA estimation: A conditional bound and its use for optimally weighted localization," *IEEE Transactions on Signal Processing*, vol. 59, no. 4, pp. 1612–1623, Apr. 2011, ISSN: 1053-587X. DOI: 10.1109/TSP.2010.2103069.

[10] R. Ulman and E. Geraniotis, "Wideband TDOA/FDOA processing using summation of short-time CAF's," *IEEE Transactions on Signal Processing*, vol. 47, no. 12, pp. 3193–3200, 1999.

[11] S. Stein, "Algorithms for ambiguity function processing," *IEEE Transactions on Acoustics, Speech, and Signal Processing*, vol. 29, no. 3, pp. 588–599, Jun. 1981, ISSN: 0096-3518. DOI: 10.1109/TASSP.1981.1163621.

[12] M. L. Fowler and X. Hu, "Signal models for TDOA/FDOA estimation," *IEEE Transactions on Aerospace and Electronic Systems*, vol. 44, no. 4, pp. 1543–1550, Oct. 2008.

About the Author

Nicholas A. O'Donoughue is a senior engineer with the RAND Corporation, where he provides radar, communications, and electronic warfare expertise to a broad array of defense studies through RAND's Federally Funded Research and Development Centers. He earned his B.Cp.E. degree from Villanova University in 2006, and both an M.S. and Ph.D. in electrical engineering from Carnegie Mellon University in 2009 and 2011, respectively. Following the completion of his Ph.D., Dr. O'Donoughue worked in the Airborne Radar Systems and Techniques Group at MIT Lincoln Laboratory from 2012–2015, where he focused on the analysis and development of electronic warfare techniques for airborne ground surveillance radar systems.

Dr. O'Donoughue was an instructor for several iterations of the "Build a Radar Course" offered through MIT's Professional Education Program from 2013–2015, and offered a course on radar signal processing at Tufts University in the spring of 2015.

He is a recipient of the National Defense Science and Engineering Graduate Fellowship, the Dean Robert D. Lynch Award from the Villanova University Engineering Alumni Society, and the Computer Engineering Outstanding Student Medallion from Villanova University. He is the author of *Emitter Detection and Geolocation for Electronic Warfare*, and has published more than 40 technical journal and conference papers. Dr. O'Donoughue is a senior member of the IEEE, and a member of both the Tau Beta Pi and Eta Kappa Nu engineering honor societies.

Index

www.ingramcontent.com/pod-product-compliance
Ingram Content Group UK Ltd.
Pitfield, Milton Keynes, MK11 3LW, UK
UKHW050300071125
464652UK00008B/13

9 781630 818883